T3-BSF-927

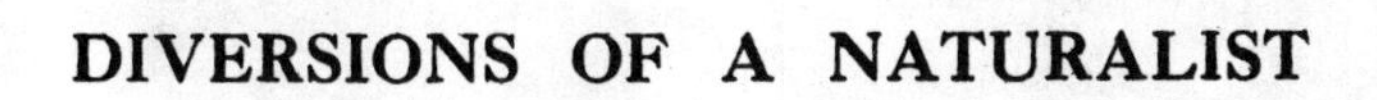
DIVERSIONS OF A NATURALIST

A CORNER IN A MARINE AQUARIUM, PAINTED BY PHILIP HENRY GOSSE, F.R.S.

DIVERSIONS OF A NATURALIST

BY

Sir EDWIN RAY LANKESTER

WITH A FRONTISPIECE AND FORTY-THREE OTHER ILLUSTRATIONS

Essay Index Reprint Series

BOOKS FOR LIBRARIES PRESS
FREEPORT, NEW YORK

First Published 1915
Reprinted 1970

STANDARD BOOK NUMBER:
8369-1471-6

LIBRARY OF CONGRESS CATALOG CARD NUMBER
77-105024

PRINTED IN THE UNITED STATES OF AMERICA

PREFACE

AT this time of stress and anxiety we all, however steadfast in giving our service to the great task in which our country is engaged, must, from time to time, seek intervals of release from the torrent of thoughts which is set going by the tremendous fact that we are fighting for our existence. To very many relief comes in splendid self-sacrificing action, in the joyful exercise of youthful strength and vigour for a noble cause. But even these, as well as those who are less fortunate, need intervals of diversion—brief change of thought and mental occupation—after which they may return to their great duties rested and refreshed.

I know that there are many who find a never-failing source of happiness in acquaintance with things belonging to that vast area of Nature which is beyond and apart from human misery, an area unseen and unsuspected by most of us and yet teeming with things of exquisite beauty; an area capable of yielding to man knowledge of inestimable value. Many are apt to think that the value of "Science" is to be measured mainly, if not exclusively, by the actual power which it has conferred on man — mechanical and electrical devices, explosives, life-saving control over disease. They would say of Science, as the ignoble proverb tells us of Honesty, that it is "the best policy." But Honesty is far more than that, and so is Science. Science has revealed to man his own origin and history, and his place in this world of un-ending marvels and beauty. It has given him a new and unassailable outlook on all things both great and small. Science commends itself to us as does Honesty and as does great Art and all fine thought and deed—not as a policy yielding material profits, but because it satisfies man's soul.

I offer these chapters to the reader as possibly affording to him, as their revision has to me, a welcome escape, when health demands it, from the immense and inexorable obsession of warfare. The several chapters have been selected from articles entitled "Science from an Easy Chair" written in recent years by me for the "Daily Telegraph." Under that title I have already published two volumes of similar selections. I have chosen a new title, "Diversions of a Naturalist," for this third volume in order to avoid confusion with the earlier ones. Illustrative drawings have been introduced into several of the articles and a few alterations made in the text. But they remain essentially what their origin implies—namely, detached essays addressed to a wide public.

I wish to thank my friend Dr. Smith Woodward of the Natural History Museum for the figures 23, 25, 26,

27, 28, 29, and 30, illustrating Chapter X, and also to thank Messrs. Veitch for the use of figures 33, 34, 35, 40, and 42. I have copied figures 4 to 8, 11, 19, and 20 from the drawings made by Philip Henry Gosse, F.R.S., and published by him in that wonderful little book "Marine Zoology," now long out of print. I have also borrowed my frontispiece from the book on "The Aquarium" by that great naturalist and lover of the seashore. Many beautiful coloured plates of marine animals executed by his skilful hand are to be found in that and other works published by him.

E. R. L.

16 *June* 1915

CONTENTS

LIST OF ILLUSTRATIONS

OH! how light and lovely the air is upon the earth! How beautiful thou art, my earth, my golden, my emerald, my sapphire earth! Who, born to thy heritage would choose to die, would wish to close his eyes upon thy serene beauties and upon thy magnificent spaces?—FEODOR SOLOGUB.

DIVERSIONS OF A NATURALIST

CHAPTER I

ON A NORWEGIAN FIORD

THE splendour of our Sussex Weald, with its shady forests and lovely gardens, around which rise the majestic Downs sweeping in long graceful curves marked by the history of our race, has charmed me during these sunny days of June. The orchids, the water-lilies, the engaging and quaintly named "petty whin," and the pink rattle are joined with the tall foxgloves and elder-blossoms in my memory. And for some reason—perhaps it is the heat—I am set thinking of very different scenes—the great, cool fiords of Norway, with their rocky islets and huge, bare mountain-tops, where many years ago I had the "time of my life" in exploring with the naturalist's dredge the coral-grown sea-bottom 1000 and even 2000 feet in a straight line below the little boat in which I and my companion and three Norwegian boatmen floated on the dark purple waves.

To let a dredge—an oblong iron frame some three feet long, to the edges of which a bag of strong netting is laced, whilst the frame is hung to a rope by a mystical triangle—sink from the side of a boat and scrape the surface of the ocean-floor far below for some ten or twenty minutes, and then to haul it up again and see what living wonders the unseen world has sent you, is, in my opinion, the most exciting and delightful sport in which a naturalist can indulge. There are difficulties and drawbacks connected with it. You cannot, in a small boat and without expenditure of large sums on a steam yacht and crew, reach from our coast—with rare exceptions in the north-west—with a fair prospect of returning in safety, those waters which are 100 fathoms deep. And it is precisely in such depths that the most interesting "hauls" are to be expected. I had had in former days to be content with 10 fathoms in the North Sea and 30 to 40 off the Channel Islands.

Then there is the question of sea-sickness. Nothing is so favourable to that diversion as slowly towing a dredge. I used to take the chance of being ill, and often suffered that for which no other joy than the hauling in of a rich dredgeful of rare sea creatures could possibly compensate, or induce me to take the risk (as I did again and again). I remember lying very ill on the deck of a slowly lurching "lugger" in a heaving sea off Guernsey, when the dredge came up, and as its contents were turned out near me, a semi-transparent, oblong, flattened thing like a small paper-knife began to hop about on the boards. It was the first specimen I ever saw alive of the "lancelet" (Amphioxus), that strange, fish-like little creature, the lowest of vertebrates. I recognized him and immediately felt restored to well-being, seized the young stranger, and placed him in a special glass jar of clear sea-water. A few years later the fishermen at Naples would bring me, without any trouble to myself, twenty or more any day of the week ("cimbarella" they called them), and I not only have helped to make out the cimbarella's anatomy, but also to discover the history of the extraordinary changes it undergoes as it grows from the egg. I sent my pupil Dr. Willey, now professor in Montreal, one summer to a nearly closed sea-lake, the "pantano" of Faro, near Messina, where the lancelet breeds. He brought home hundreds of minute young in various stages, and again later made a second visit to that remote sea-lake in order to complete our knowledge of their growth and structure by observation on the spot.

The advantage of the Norwegian fiords for a naturalist who loves to "dredge" is that at many parts of the coast you can sail into water of 200 fathoms depth and more, within three minutes from the rocky shore; and, secondly, that the great passage between the islands and the mainland is, to a very large extent, protected from those movements of the surface which cause such torture to many innocent people who venture on the sea in boats! Accordingly, in 1882, when I heard from the greatest naturalist-dredger of his day—the Rev. Canon Norman, of Durham—that he knew a farmhouse at Lervik, on the island of Stordö, near the mouth of the Hardanger Fiord, between Bergen and Stavanger—where one could stay, and where a boat could be hired for a couple of months—I determined to go there. I was confirmed in my purpose by the fact that Canon Norman had obtained in his dredge, at a spot near Lervik, which he marked for me on the large-scale official map of the region, a very curious little polyp-like animal, attached to and branching on the stems of the white coral which one dredges there at the depth of 150 fathoms. The little animal in quest of which I went, though other wonderful things were to be expected also, had been dredged originally by Dr. Norman off the Shetland Islands, and described by

Professor Allman, of Edinburgh. But they had not examined it in the living state with the microscope, and though they showed that it was quite unlike other polyps, yet there was obvious need for further examination of it. I hoped to obtain its eggs and to watch its early growth. The name given to it by Allman was "Rhabdopleura," meaning "rod-walled," alluding to a rod-like cord which runs along the inside of the delicate branching tube (only the one-twentieth of an inch wide), which the little animal constructs and inhabits.

I sent a chest containing glass jars, microscopes, books, chemicals, etc., and my dredge, as well as a large windlass, on which was coiled 600 fathoms of rope, by sea to Lervik, and started in early July, with my assistant, Dr. Bourne (afterwards Director of Education in the Madras Presidency), overland, via Copenhagen, for Christiania. Thence we drove in "carioles" across Norway to Laerdalsören, on the west coast, making acquaintance with the magnificent waters—rivers, lakes, and cascades—of that pine-grown land. After visiting the Naerodal and the glaciers which descend from the mountains into the sea on the Fjaerlands Fiord, we took steamer to Lervik, and were welcomed at our farmhouse by its owner, the sister of the member of Parliament for the surrounding region (about four times the area of Yorkshire), whose son secured for me a fair-sized sailing boat, and with two other men of Lervik engaged as my crew for six weeks.

After a day or two we had everything in order, and at seven o'clock one morning sailed out of the harbour to make our first cast of the dredge. The mouth of the harbour of Lervik is 40 fathoms deep, and the great north-bound steamers enter it and come alongside the

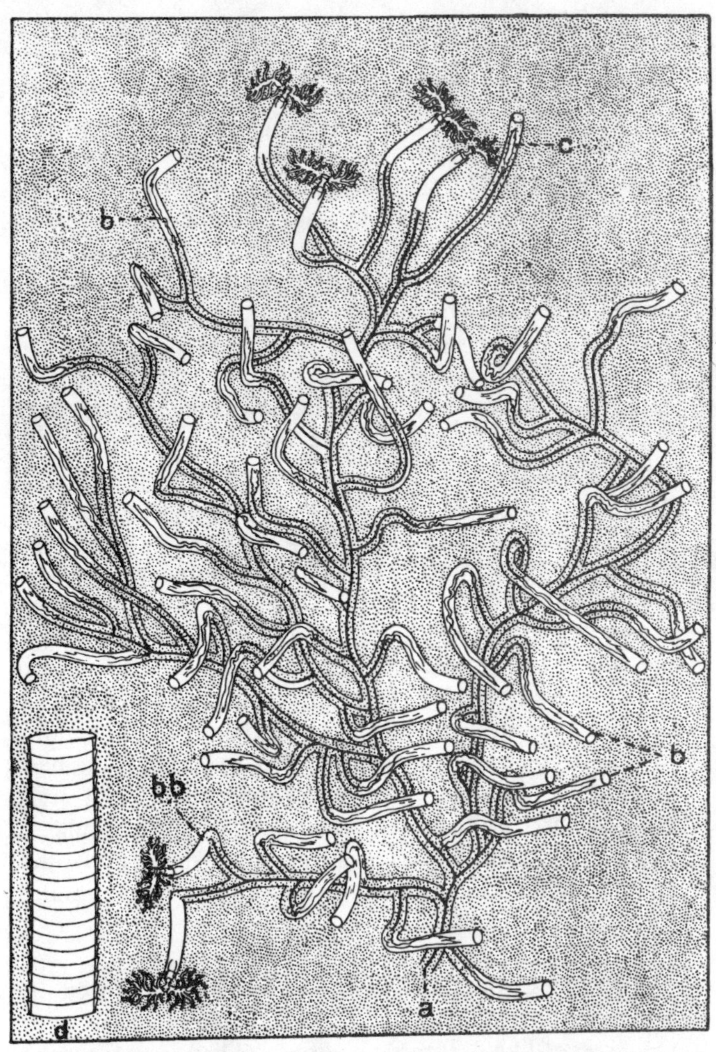

FIG. 1.—A portion of the branching tubular growth formed by Rhabdopleura Normani, fixed to and spreading over the smooth surface of an Ascidian, dredged at Lervik and drawn of three times the natural dimensions. The colourless tubes (b) stand up freely from the surface to which the rest of the growth is adherent, and from each of them issues in life (as seen at bb) a polyp such as that shown in Fig. 2. Each polyp is continuous with the dark internal cord (or rod) which is seen traversing the whole of the tubular system. a, points to the main and oldest portion of the branching stem; c, points to a "leading" shoot which is still adherent and will give rise to young buds right and left which will form upright tubes like b. The inset d represents a piece of the tube magnified so as to show the rings by which it is built up.

rocks on which the village stands. Outside the harbour the depth increases precipitously to 200 fathoms. We sailed about 10 miles along the fiord, and determined precisely the spot indicated by Dr. Norman on the map, and here we lowered our dredge. We had fixed around the mouth of the dredge long tassels of hemp fibre, since on rocky ground, such as we were now dredging, one cannot expect much to be "scooped up" by the slowly travelling dredge as it passes over the bottom, whilst the threads of the hemp, on the contrary, entangle and hold all sorts of objects with which they come into contact. We were 1000 feet from the bottom, and our dredge took a good five minutes to sink as we paid out the rope from the winch in the stern of our boat. When it reached the bottom we let out another 2000 feet of rope, and then very slowly towed the dredge for about a quarter of an hour. Then the laborious task commenced of winding it up again, two men turning the handles of the winch for a quarter of an hour. At last the dredge could be seen through the clear water, and soon was at the surface and lifted into the boat. The hempen tangles were crowded with masses of living and dead white coral (Fig. 3), star-fishes, worms, and bits of stone covered with brilliant-coloured sponges, Terebratulæ (a deep-water, peculiar shellfish, the lamp-shell), and other animals. There were only a few fragments of coral in the bag of the dredge.

We filled glass jars with sea water and placed the bits of coral in them, and I eagerly examined them for the creeper-like "Rhabdopleura." There, sure enough, it was on several of the dead stems of coral, and we sailed back to Lervik with our booty in order to examine it at leisure with the microscope whilst still fresh and living. In our temporary laboratory at the farmhouse

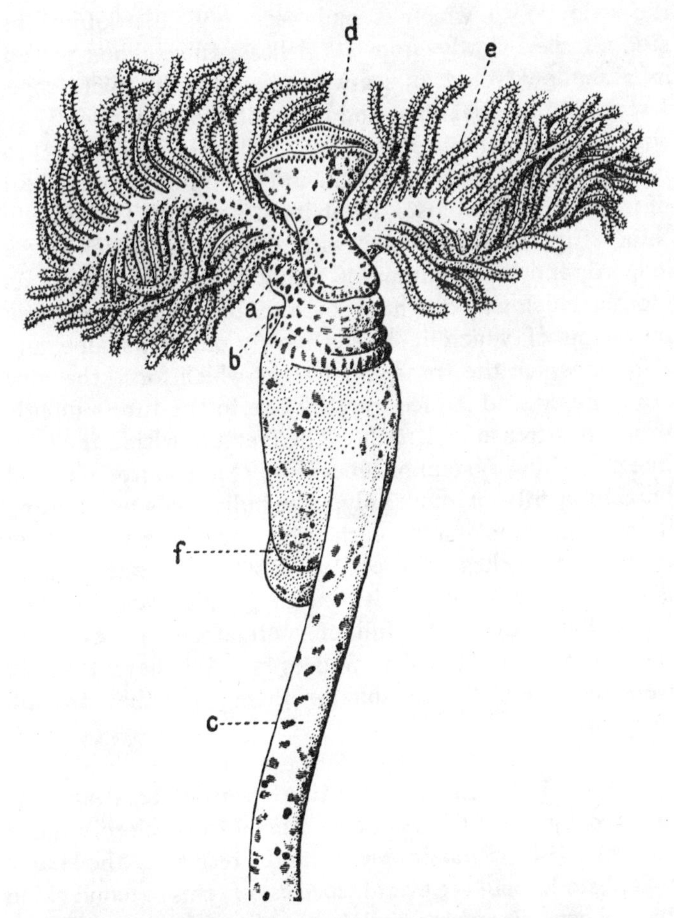

FIG. 2.—One of the polyps of Rhabdopleura which is attached by its soft contractile stalk (c) to the dark internal cord seen in Fig. 1. A similar polyp issues during life from the open end of each of the upright tubes seen in Fig. 1, and is, when disturbed, pulled back into the tube by the contraction of the cord c. a, mouth; b, vent; c, contractile stalk; d, head-shield or disk; e, the left gill-plane; f, the body-mass enclosing the intestine, etc. (From a drawing made by the author in Lervik, Stordö, in 1882.) For a full account of Rhabdopleura, see the "Quart. Journal of Microscopical Science," vol. xxiv., 1884.

the little polyp which it had been my chief object to study, issued slowly from its delicate tubes when placed in a shallow trough of sea-water beneath the microscope. I was able on that day, and many others subsequently—with renewed supplies from the depths of the fiord—to make coloured drawings of it, and to find out a great deal of interest to zoologists about its structure. The minute thing (Fig. 2) was spotted with orange and black like a leopard, and had a plume of tentacles on each side of its mouth, which was overhung by a mobile disk—the organ by means of which it creeps slowly out of its tube, and also by which the transparent rings which form the tube are secreted and added one by one to the tube's mouth, so as to increase its length. The creature within the tree-like branching system of tubes (Fig. 1) is also tree-like and branching, fifty or more polyp-like individuals terminating its branches and issuing each from one of the upstanding terminal branches of the tube system. I was able to determine the "law" of its budding and branching, and I also found the testis full of spermatozoa in several of the polyps, but I failed to find eggs. I believe that we were too late in the season for them; and they are still unknown.

One of the most interesting deep-sea creatures discovered by the "Challenger" proved to be closely allied to our little Rhabdopleura, and received the name "Cephalodiscus." Several species of this second kind have been discovered in the last twenty years in the deep sea, and the largest and most remarkable in some respects was one which "jumped to my eyes" among the booty of marine dredgings sent home from the Antarctic expedition of the "Discovery" by Captain Scott, when I unpacked the cases containing these marine treasures, in the basement of the Natural History

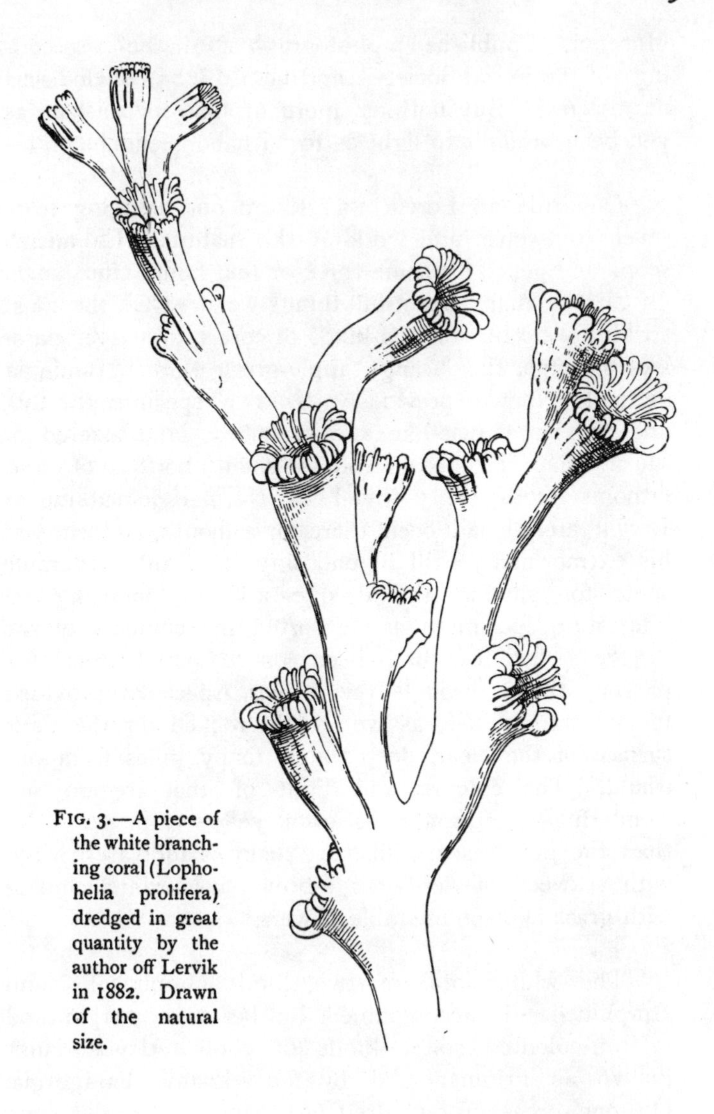

FIG. 3.—A piece of the white branching coral (Lophohelia prolifera) dredged in great quantity by the author off Lervik in 1882. Drawn of the natural size.

Museum. I published a photograph of it in the "Proceedings of the Royal Society," and named it "Cephalodiscus nigrescens." But nothing more of importance has, as yet, been brought to light as to "Rhabdopleura."

Our rule at Lervik was to go out dredging from seven to twelve, and work at the material with microscope and pencil for some three or four hours after lunch. Of all the many beautiful things we dredged, the most striking were the various kinds of corals, the large, glass-like shrimps, the strange apple-green worm Hamingia (actually known previously by two specimens only), and the large, disc-like and branched, sand-covered or sausage-like Protozoa (from a shelly bottom of 200 fathoms depth). My friend Dr. Norman joined me at Lervik after I had been there for a month, and showed his extraordinary skill in choosing the most favourable spots for sinking the dredge and in pouncing on interesting specimens as we sorted the contents of the dredge (when we had been on a soft bottom) by passing them through the sieves, specially provided for naturalists' use, as we gently rocked on the dark surface of the clear, deep water, many miles from our island. The colours and light of that region are wonderful—the mountains of a yellow tint, far paler than the purple sea, whilst the rocky islands are fringed with seaweed of rich orange-brown colour, and clothed with grass and innumerable flowers.

The white coral of two kinds (Lophohelia and Amphihelia) is accompanied by beautiful purple and salmon-coloured softer kinds of coral (Alcyonarians), known as Primnoa, and by the gigantic Paragorgia. On one occasion our dredge became fast. For long nothing would move it, and we feared we should have to cut it and lose some 300 fathoms of rope. At last the efforts of four men at the oars set it free, and we wound it in. As the dredge came up we found entangled in the rope an enormous tree-like growth, as thick as a man's arm, seven feet long, and spreading out into branches, the whole of a pale vermilion colour (like pink lacquer)—a magnificent sight! It was a branch of the great tree-coral of these waters—the Paragorgia—and we preserved many pieces of it in alcohol and dried the rest. But the gorgeous colour could not be retained.

One day the green worm, Hamingia (named after a Norwegian hero—Haming) was dredged by us at the mouth of Lervik Harbour, in 40 fathoms. A somewhat similar worm lives in holes in the limestone rocks of the Mediterranean, and is named Bonellia (after the Italian naturalist, Bonelli). All the specimens of this Mediterranean worm, which is as large as a big walnut, and has a trunk, or proboscis, a foot long, were found to be females. The male was unknown until my friend the late Alexander Kowalewsky, the most remarkable of Russian zoologists, discovered that it is a tiny threadlike green creature, no bigger than the letter "i" on this page. Three or four are found crawling about on the body of the large female. I found the same diminutive kind of male crawling on my Norwegian Hamingia, at Lervik, and published a drawing and description of him. I was also able to show that, unlike Bonellia, the Norwegian worm has red blood-corpuscles, like those of a frog, and impregnated with hæmoglobin, the same oxygen-carrying substance which colours our own blood-corpuscles. The identity of the worm's hæmoglobin with that in our own blood was proved by its causing two dark bands of absorption in the solar spectrum when light was passed through it and then through the spectroscope—dark bands exactly the same in position and intensity as those caused by the red substance of my own blood and changing into one single band intermediate in position between the two—when deprived by an appropriate chemical of the oxygen loosely combined with it.

On the Fiord near Lervik.

Of many other things we caught and many other delights of that long-past summer on the Norwegian fiords, of the great waterfalls, the vast forests, the delightful swimming in the sea, the trout-fishing, and the very trying food approved and provided for us by the natives, I must not now tell. My hope is that I may have enabled my readers to understand some of the enjoyment open to the marine zoologist, even when he dispenses with the aid of a big steamship, and modestly pursues his quarry in a sportsmanlike spirit.

CHAPTER II

NATURE-RESERVES

ONE of the new features of modern life—the result of the enormous development of the newspaper press and the vast increase in numbers of those who read and think in common—is the development of a sensitive "self-consciousness" of the community, a more or less successful effort to know its own history, to value the records of the past, and to question its own hitherto unconscious, unreflecting attitude in mechanically and as it were blindly destroying everything which gets in the way of that industrial and commercial activity which is regarded, erroneously, as identical with "progress." Beautiful old houses and strange buildings—priceless records of the ways and thought of our early ancestors—which at one time were either guarded by superstitious reverence or let alone because there was room for them and for everything else in the spacious country-side—have been thoughtlessly pulled down as population and grasping enterprise increased. The really graceful old houses of London and other towns, lovingly produced by former men who were true artists, have been broken up and their panelling and chimney-pieces sold to foreigners in order to make way for more commodious buildings, hideous in their ignorant decoration, or brutally "run up," gaunt, bare, and mis-shapen. The stones of Avebury, of Stonehenge, and of many another temple have been knocked to pieces by emancipated country-folk—no longer restrained either by superstition or by reverence—to mend roads and to make enclosures.

Happily the new self-consciousness is taking note of these things. That strange lumbering body which we call "the mother of parliaments" has dimly reflected the better thought of the community, and given a feeble sort of protection to ancient monuments. The newspapers have lately managed to excite some public interest in a fine old house in Dean Street, Soho, and to arouse a feeling of shame that the richest city in the richest Empire of the world should allow the few remnants of beautiful things of the past still existing in its midst to be destroyed by the uncontrolled operation of mercenary "progress." I have, in common with many others, visited this doomed mansion. It is a charming old place, of no great size or importance, and, with its well-proportioned panelled rooms and fine staircase, was destined to be a private residence. It is not large enough to be a museum, but its rooms might serve for the show place of a first-rate maker or vender of things of fine workmanship. There ought to be some public authority—municipal or departmental—with power to acquire such interesting houses as this, not necessarily to convert them into permanent public shows, but to keep them in repair, and to let them on lease, at a reasonable rent, to tenants, subject to the condition of their being open on certain days in the year to artists and others provided with orders of admission by the authority. In other countries such arrangements are made; with us they are not made simply because we have not assigned to any authority the duty of acting in this way for the public benefit. Our public authorities have little or no public spirit, and resemble private committees, councils, and individuals in evading and refusing even the smallest increase of responsibility and activity beyond that which they are compelled by law to discharge. Unless they are legally compelled to interfere, all records of art and nature may perish before they will incur the inconvenience of moving a finger! Consequently the only thing to be done is to assign such duties by law to an existing authority, or to one created for such purposes.

The same tale of destruction and irreparable damage has to be told of our dealings with the beauty of once unsullied moorland, meadow, marsh, forest, river-bank, and seashore. But the destruction has here been more gradual, less obvious on account of remoteness, and more subtle in its creeping, insinuating method, like that of a slowly-spreading infective disease. The word "country" has to a very large extent ceased to signify to us "outlying nature beyond the man-made town," occupied only in little tracts here and there by the immemorial tillers of the soil. The splendid and age-long industry of our field-workers has made much of our land a garden. Now they themselves are disappearing or changed beyond recognition, losing their traditional arts and crafts, their distinctive and venerable dialects, and their individuality. The land is enclosed, drained, manured; food plants produced by the agriculturist replace the native plants; forests are cut down and converted into parks and pheasant-runs; foreign trees are substituted for those native to the soil. Commons, heaths, and wild moorlands have been enclosed by eager land-grabbers, the streams are polluted by mining or chemical works, or if kept clean are artificially overstocked with hand-fed trout; whilst the open roads reek of tar and petroleum. The "wilderness" is fast disappearing, and it is by this name that we must distinguish from the mere "country," as much besmirched and devastated by man as are the sites of his towns and cities, the regions where untouched nature still survives and is free from the depredations of humanity. Many beautiful and rare plants which once inhabited our countryside have perished; many larger animals (such as wolf, beaver, red-deer, marten-cats, and wild-cats) have disappeared, as well as many insects, great and small, such as the swallow-tailed butterfly and the larger copper butterfly, and many splendid birds.

Here and there in these islands are to be found bits of "wilderness" where some of the ancient life—now so rapidly being destroyed—still flourishes. There are some coast-side marshes, there are East Anglian fens, some open heath-land, and some bits of forest which are yet unspoilt, unravaged by blighting, reckless humanity. It is a distressing fact that some of the recent official attempts to preserve open forest land and commons for the public enjoyment have been accompanied by a mistaken attempt to drain them, and lay them out with gravel walks, to the complete destruction of their natural beauty and interest. The bog above the Leg of Mutton Pond, on Hampstead Heath, where I used to visit, years ago, the bog-bean and the sun-dew, and many a moss-grown pool swarming with rare animalcules, has been drained by an over-zealous board of guardians, animated by a suburban enthusiasm for turf and gravel paths. The same spirit, hostile to nature and eager to reduce the wilderness to vulgar conventionality, has tamed the finer parts of Wimbledon Common, and is busy laying down gravel paths in Epping Forest. In the New Forest the clamour of the neighbouring residents for "sport" has led to the framing of regulations by the officials of the Crown (it is a "Royal" forest), which are resulting in the destruction and disappearance of rare birds which formerly nested there. Many a distant common threatened by the builder has been preserved as an open space by golfers. Such preservation is like that of the boards of conservators, useless from the point of view of the nature-lover. The health-seeking crowd spreads devastation around it. The rare sand-loving plants of the dunes, and the "bog-bean," the "sun-dew," and other refugees from human persecution on our once unfrequented heath-lands, are remorselessly trodden down or hacked up by the golfer. Other destroyers of nature's rarer products are those who greedily search for them and carry them off, root and branch, to the last specimen, in order to sell them. These dealers are "collectors," indeed, but must not be confused with the genuine "naturalist," who may allow himself, with due modesty, to secure a limited sample of treasures from nature's open hand.

Under these circumstances a society has been founded for the formation of "nature-reserves" in the British Islands. Its object is to secure, by purchase or gift, tracts of as yet unsullied wilderness—of which some are still, though rarely, to be found—where beast and bird, insect and plant are still living as of old—untouched, unmolested, undisturbed by intrusive, murderous man. The society's object is to enter into relations with those who may know of such tracts, and to arrange for their transference—if of sufficient interest—to the National Trust. The expense of proper guardianship and the admission to the reserve of duly authorized persons would be the business of the society. Its office is at the Natural History Museum in Cromwell Road, and Mr. Ogilvie Grant, the naturalist in charge of the ornithological collections, is one of the secretaries. Sir Edward

Grey and Mr. Lewis Harcourt and several of our most distinguished botanists and zoologists are members of the council. All who sympathize with the objects of the society should write to the secretary for further information.

Already two tracts of land were secured as nature-reserves before the society came into existence. One of these is Wickham Fen, not far from Cambridge, renowned for its remarkable plants and insects. It was purchased and placed in the hands of the National Trust by a public-spirited entomologist. Another reserve, which has been secured, is far away on the links or dunes of the north coast of Norfolk, and is of especial interest to botanists. No one—either golfer or bungalow-builder—can now interfere there and destroy the interwoven flora and fauna, the members of which balance and protect, encourage and check one another, as is Nature's method. The interaction of the various species of wild plants in this undisturbed spot is made the subject of continual and careful study by the botanists who are permitted to frequent it. More such "reserves" and of different characters are desirable. Should we, of the present day, succeed in securing some great marsh-land, one or more rocky headlands or islands, and a good sweep of Scotch moor and mountain, and in raising money to provide guardians for these acquisitions, we shall not only enjoy them ourselves but be blessed by future generations of men for having saved something of Britain's ancient nature, when all else, which is not city, will have become manure, shooting greens, and pleasure gardens.

In Germany and in Switzerland a good deal has been done in this way. Owing to the existence of "forestry" and a State Forest Department in Germany—which has no representative in this country—there is machinery for selecting and guarding such "reserves." A large sum is assigned annually by the Government to this purpose. Last year an international congress, attended by delegates from the English society, as well as by representatives of many other States, was held, and much useful discussion as to methods and results took place.

The notion of creating a nature-reserve on a small scale seems to have originated with Charles Waterton, the traveller and naturalist, who in the middle of last century converted the estate surrounding his residence near Pontefract in Yorkshire into a sort of sanctuary, where he made it a strict rule that no wild thing should be molested. For some years now the attempt to create "nature-reserves," on a far larger scale than those of which I have been writing, has been made where civilization is planting its first settlements in primeval forest and prairie. The United States Government, impressed with the rapid destruction and disappearance both of forests and of native animals which have accompanied the opening up by road and rail of vast territories in the West, created in 1872 the national "reserve," called the Yellowstone Park, which is some 3300 square miles in area. We are assured that here under proper guardianship the larger native animals are increasing in number; whilst the great coniferous trees, which were in danger of extermination by the white man, are safe. Similar reserves have been proclaimed in parts of Africa under British control, but though that known as Mount Elgon—an ancient volcanic cup, clad with forest, and ten miles in diameter—seems to have been effective, and to have furnished in Sir Harry

Johnston's time, ten years ago, a refuge for the giraffe, it is scarcely possible, at present, to provide an efficient police force to protect areas of something like 1000 square miles against the depredations of native and commercial "hunters" provided with modern rifles.

In May, 1900, I was, with the late Sir Clement Hill, appointed "plenipotentiary" by her Majesty Queen Victoria to meet representatives of Germany, France, Spain, Portugal, and the Congo States in a conference, presided over by the late Marquis of Linlithgow, at the Foreign Office. The conference was arranged by the great African powers in order to consider and report on the means to be taken to preserve the big game animals of Africa from extinction. We spent an extremely interesting fortnight, and finally agreed upon a report, the upshot of which was that whilst certain animals, such as the giraffe, some zebras and antelopes, the gorilla, and such useful birds as the vultures, secretary bird, owls, and the cow-pickers (Buphagus), should be absolutely protected, others should be only protected at certain seasons, or in youth, or in limited numbers, and others again should be killed without licence or restraint at any time, such being the lion, the leopard, the hunting-dog, destructive baboons, most birds of prey, crocodiles, pythons, and poisonous snakes. The question of large "nature-reserves" was discussed. It was agreed that such reserves should be maintained for the breeding-places and rearing of the young of desirable animals, and that the destruction of predatory animals or an excess of other forms should be permitted to the administrators of such reserves. Thus it is clear that no absolute "nature-reserves" were considered possible.

In fact this is the case whether the reserve be large

or small. Once man is present in the neighbourhood, even at a long distance, he upsets the "balance of Nature." The naturalist's small "nature-reserve" may be ravaged by predatory animals driven from the outlying region occupied by man, or again, the absence from the "reserve" of predatory animals which act as natural checks on the increase of other animals, may lead to excessive and unhealthy multiplication of the latter. Man must "weed" and artificially manage his "reserve" after all! Man brings also into the neighbourhood of reserves, great and small, disease germs in his domesticated animals, which are carried by insects into the cherished "reserve," and there cause destruction. Conversely, the animals maintained in a reserve carry in their blood microscopic parasites to the poisons of which they have become immune by natural selection in the course of ages. They act as "reservoirs" of such microscopic germs. These germs carried by flies or other insects to the carefully reared cattle imported by civilized man from other regions of the world into the neighbourhood of such "reserves," cause deadly disease (such as the tsetse-fly diseases or trypanosome diseases) to those imported cattle, as also to man himself. Whilst, then, we may do something to retain small tracts of our own country in the modified state which it attained after the earlier inhabitants had destroyed lion, bear, wolf, and other noxious animals, as well as great herbivora, such as giant deer, red deer, aurochs (or great bull), and bison—yet in reality a true "Nature-reserve" is not compatible with the occupation of the land, within some hundreds of miles of it, by civilized, or even semi-civilized, man.

Nothing but the isolation given by a wide sea or high mountain ranges will preserve a primeval fauna and flora—the indigenous man-free living denizens of the isolated region—from destruction by the necessary unpremeditated disturbance of Nature's balance by man once he has passed from the lowest stage of savagery. At present we are faced by this difficulty in Africa. Not only the white settlers have large herds of cattle, but before their arrival the native races had imported Indian cattle. These cattle are destroyed by "fly disease," the germs (trypanosomes) being carried by the tsetse fly to the domesticated cattle from wild buffalo which swarm with the germs but are uninjured by them. Consequently, if the rich pasture lands of Africa—at present unutilized—are to be occupied by herdsmen, the wild game, buffalo and antelopes, must be destroyed. In many regions they have been destroyed. Is this destruction to be continued? If Africa is to be the seat of a modern human population and supply food to other parts of the world, the whole "balance of Nature" there must be upset and the big wild animals destroyed. There is no alternative. The practical question is, "How far is it possible to mitigate this process?" Can a great African "reserve" of 100,000 square miles be established in a position so isolated that it shall not be a source of disease and danger to the herdsmen and agriculturists of adjacent territory?

CHAPTER III

FAR FROM THE MADDING CROWD

SOME men of unbalanced minds have lately proposed deliberately and completely to obliterate all the artistic work of past generations of man in order, as they openly profess, that they themselves and their own productions may obtain consideration. Even were they able to make such a clearance, it may be doubted whether the consideration given to their own performances would be favourable. These obscure individuals have immodestly dubbed themselves "futurists," and the name has been at once adopted as a mystification and advertisement by a variety of art-posers—probably unknown to the originators of the word—who have ventured into one or other of the fields of art without even the smallest gift, either of conception or of expression, or even of imitation. They receive undeserved attention from a section of the public ready to dabble in every newly-made puddle. I am led to refer to them because the abolition of the supremely beautiful things slowly evolved by Nature in the long course of ages, and the substitution for them of man's fancy breeds and races and garden paths, is not merely a parallel piece of folly, but is due to a mental defect identical with that of the genuine "futurist," namely, an intellectual incapacity which renders its victim insensible to the charm of historical and evolutional complexity.

The modern man who nourishes a real love for undistorted nature—that is to say, who is a true "naturalist"—has one or two resources even in these British Islands. There are ways of access to Nature unadorned by man which are open even to the town-dweller. The chief of these is the seashore. Even from London, in the course of a few hours, one may be transported to territory where there are no traces of man's operations. The region of rock and pool, sand-flat, and shell-bank, exposed by the sea as it retreats, is a real "nature-reserve"—effectually so is that deepest area only exposed at spring-tides. The locality chosen by the naturalist must be at a distance from any great harbour or estuary polluted by the cities seated on its banks, and should also be out of the way of the modern steam-driven fish trawlers, which have caused havoc in some sweet bays of our southern coast by pouring out tons of dead, unsaleable fish. The rejected offal has become the gathering-ground of carnivorous marine creatures, and the balance of Nature has been upset by the nourishment thus thoughtlessly thrown by man into new relations.

Some favoured spot on the south or west coast may be known to our city-dwelling nature-lover, and thither he will hasten to spend week-ends, and, when he can, longer spells in the supreme delight of undisturbed communion with the things of Nature, apart from human "enterprise." In some cottage near the sea marsh, where an unpolluted stream joins the salt water, he has his accustomed lodging; his host, a cheery long-shore fisherman and handy boatman. Close by is the rising headland and rocky cliff facing the sea. The shore is strewn with rocks, and as the tide goes down long "reefs" are exposed, clothed with brown and green seaweeds. Here no man has intruded! When the water recedes still farther, pools and miniature caverns appear, edged with delicate feathery red-coloured seaweeds. Many small fishes, shrimps of various kinds, sometimes pale rainbow-tinted "squids" (one of the more delicate cuttle-fishes), are seen darting about the pools, changing their colour with lightning rapidity. The overhanging sides of the rock-pools give protection to gorgeously-coloured "sea-anemones" adhering to them. Here, also, are those exquisite ascidians—ill-described by the rough name "sea-squirt"—hanging from the rocks like drops of purest crystal in their transparency —for which naturalists use the prettier title "Clavellina." The nature-lover now turns one of the large flat slabs of rock lying in such a pool—well knowing what loveliness its under-side will reveal to his eyes. That under-side is studded with a dozen or two of the most exquisite gems of green and peach colour, ruby and yellow (Corynactis by name!), which, if the slab of stone is left beneath the water, expand and display each its circlet of brilliant little tentacles. They are sea-anemones no bigger than the precious stone in a signet-ring. Among them a bright salmon-coloured worm hastens with serpentine movement and the rippling strokes of a hundred feathery feet to escape from the unaccustomed light. A deep blood-red coloured prawn (Alpheus) darts from concealment and hastily buries itself in the sandy bottom of the pool, snapping its pincerlike claw with a sharp cracking sound. A couple of bivalved shells (Lima hians) which were concealed beneath the slab swim lazily round the pool by opening and closing their delicate white "valves"—an unusual kind of activity in such mussels, oysters, and clams—whilst a fringe of long orange-red tentacles trails in the water from each of them. The lifting of another rock may dislodge an "octopus"— or a huge brilliantly-coloured star-fish—or one of the rarer kinds of crab eager to avoid the observation of the octopus, of which it is the regular food. A spade pushed into the neighbouring sandbanks reveals heart-urchins, gorgeous sea-worms, and burrowing shell-fish and perhaps sand-eels. The human visitor—bending over these scenes of wonderment and perhaps venturing to transfer one or two only of the less familiar animals to a glass jar filled with sea-water so that he may see them more clearly—at last stands up and straightens his back, gazing over the sun-bathed scene from the tumbled weed-grown rocks, encrusted with crowds of purple-blue mussels, to the patches of golden sand, clear pools, and the blue sea beyond. Then he may note (as I have) a curious rhythmical sound if he is among rocks covered with sea-weeds—a quiet but incessant "hiss-hiss," which is heard above the deeper-toned lapping of the little waves among the big stones. This is the sound made by the rasp-like tongues of the periwinkles feeding on the abundant weed, over which they crawl, leaving the water and "browsing" on the surface exposed to the air by the fall of the tide. The browsing sound of these little snails is to the sea-shore what the humming of bees is to inland meadows.

Day after day and at various seasons of the year the nature-lover will visit this sanctuary, and, whilst contemplating the lovely forms, colour, and movement of its denizens, will learn the secrets of their life, of their comings and goings, and the mysteries of their reproduction, their birth, and their childhood. Each day he finds something unknown to his brother naturalists. He will examine it with his lens, paint it in all its beauty, and tell of it in due course in printed page and coloured portraiture; but he is no mere seeker for novelty, nor is the credit of discovery the motive of his devotion. Beyond and greater than any such gains are the incomparable delight, the never-failing happiness which personal intimacy with the secret things of natural beauty bring to him.

He has yet another chance of such enjoyment, if he be a microscopist, and familiar with the inhabitants of fresh-water ponds. A pond is, in many cases, an oasis in the waste of civilization, a miniature nature-reserve, rarely, if ever, affected by human proceedings until haply it is abolished altogether. A fairly deep, stagnant pond under trees in some secluded park is one of the most favourable kind, but all sorts deserve inquiry (even the rain pools on the roofs of old houses in Paris have rewarded the faithful seeker), and may prove, for a time at least, havens of refuge for a wonderful assemblage of animalcules and minute microscopic plants, which for the most part perish as did the bison of the American plains by the mere disturbance caused by the propinquity of civilized man. I knew such a pond—it is now built over—near Hampstead. As one lay on the bank and peered into the depths of the pond the transparent, glass-like larvæ of the "plume fly" (Corethra) could be seen swimming in the clear water, driving before it troops of minute pink-coloured water-fleas (Daphnia) and other crustaceans.

In other parts the water was made bluish-green by crowds of the little floating spherical animalcules called "Volvox globator." The mud contained many curious worms allied to the earth-worm, whilst coiled round fallen twigs were the small snake-like worms known as "Nais serpentina." Desmids, Diatoms, and animalcules of endless variety abounded. A muslin net set on a ring on the end of a stick enabled one to procure samples of the floating life of the water and also to skim the surface of the mud, and these spoils were brought home in bottles and searched for hours drop by drop with the microscope. The world of active, graceful, bustling life thus revealed as one gazes for hours through the magic tube of the microscope, is as remote from human civilization as that uncovered at low tide on the seashore. Many a worried City man, amongst them a great political writer on the staff of a London daily, now passed from among us, has found in this microscopic world—so readily accessible even at his own study table—a release from care, a refreshing contact with unadulterated natural things of life and beauty. My friend, Iwan Müller, the writer referred to, was as discriminating a judge of the shapes of wheel-animalcules as he was of the faces of the politicians of Europe and South Africa!

There is another and much more difficult escape from the grip and taint of civilization, which is that effected by the explorer who penetrates into sparsely inhabited wilds such as those of the Australian continent. Man is there, but in such small number (one to every 450 square miles!), and in so primitive and child-like a state, that he is not a disturbing element, but simply one of the "fauna"—one of the curious animals living there under the domination of Nature—not yet "Nature's rebel," but submissive, unconscious, and a more fascinating study for us than any other of her products. He shows us what manner of men were our own remote ancestors. The hunters who have left their flint implements in the earlier river gravels of Western Europe were such men as these Australian natives now are. Naked, using only sticks and chipped stones as implements and weapons, destitute of crops or herds or habitations, wandering from place to place in keen search of food—small animals, birds, lizards, and grubs

—these Australians have none of the arts of the most primitive among other races, excepting that they can make fire and construct a canoe of the bark of trees. They have not even the bow and arrow, but make use of spears and the wonderful "boomerang" in hunting and fighting. They daub themselves with a sort of white paint, and decorate their bodies with great scars made by cutting gashes in the flesh with sharp stones, and they dress their heads and faces and ceremonial wands with wool and feathers, which they fix by the aid of an adhesive fluid always ready to hand—namely, their own blood. I recently was present at a lecture given to the Anthropological Institute in London by Professor Baldwin Spencer, of Melbourne, with whom I was closely associated when he was a student at Oxford thirty years ago. He has devoted many years to the study of the Australian natives, and ten years ago published a most valuable work describing his experiences amongst them, to which he has recently added a further volume. He has lived with them in friendship and intimacy in the remote wilderness of the Australian bush, and has been admitted as a member of one of their mysterious clans, of which the "totem," or supposed spirit-ancestor, is "the witchety grub"—a kind of caterpillar. He has been freely admitted to their secret ceremonies as well as to their more public "corroborees" or dances, and has been able (as no one else has been), without annoyance or offence to them, to take a great number of cinema-films of them in their various dances or when cooking in camp or paddling and upsetting their canoes, and climbing back again from the river. Many of these he exhibited to us, and we found ourselves among moving crowds of these slim-legged, beautifully-shaped wild men. The film presented some of their strange elaborate dances, which soon will be danced no more. These wild men die out when civilized man comes near them. It appears that they really spend most of their time in dancing when not looking for food or chipping stone implements, and that their dances are essentially plays (like those of little children in Europe), the acting of traditional stories relating the history of their venerated animal "totem," which often last for three weeks at a time! Whilst dancing and gesticulating they are chanting and singing without cessation, often repeating the same words over and over again. Here, indeed, we have the primitive human art, the emotional expression from which, in more advanced races, music, drama, dancing, and decorative handicraft have developed as separate "arts."

The most remarkable and impressive result was obtained when Professor Baldwin Spencer turned on his phonograph records whilst the wild men danced in the film picture. Then we heard the actual voices of these survivors of prehistoric days—shouting at us in weird cadences, imitating the cry of birds, and accompanied by the booming of the bull-roarer (a piece of wood attached to a string, and swung rapidly round by the performer). A defect, and at the same time a special merit, of the cinema show of the present day is the deadly silence of both the performers and the spectators. Screams and oaths are delivered in silence; pistols are fired without a sound. One can concentrate one's observation on the facial expression and movements of the actors with undivided attention and with no fear of startling detonations. And very bad they almost invariably are, except in films made by the great French producers. On the other hand, I was astonished at the intensity of the impression produced by hearing the actual voices of those Australian wild men as they danced in rhythm with their songs. To hear is a greater means of revelation than to see. One feels even closer to those Australian natives as their strange words and songs issue from imprisonment in the phonograph, than when one sees them in the film pictures actually beating time with feet and hands and imitating the movements of animals. To receive, as one sits in a London lecture-room, the veritable appeal of these remote and inaccessible things to both the eye and the ear simultaneously, is indeed the most thrilling experience I can remember. With a feeling of awe, almost of terror, we recognize as we gaze at and listen to the records brought home by Professor Baldwin Spencer that we are intruding into a vast and primitive Nature-reserve where even humanity itself is still in the state of childhood—submissive to the great mother, without the desire to destroy her control or the power to substitute man's handiwork for hers.

CHAPTER IV

THE GREAT GREY SEAL

IT is always pleasing to find that intelligent care can be brought to bear on the preservation of the rare and interesting animals which still inhabit parts of these British Islands, though it is not often that such care is actually exercised. Mr. Lyell (a nephew of the great geologist Sir Charles Lyell) in April 1914 introduced a Bill into the House of Commons which is called the Grey Seals (Protection) Bill. It came on for consideration before the Standing Committee, was ordered to be reported to the House without amendment, and has now passed into law.

The Great Grey Seal is a much bigger animal than the Common Seal, the two species being the only seals which can be properly called "British" at the present day, though occasionally the Harp Seal, or Greenland Seal, and the Bladder-nosed Seal are seen in British waters, and may emerge from those waters on to rocky shores or lonely sandbanks. The Great Grey Seal is called "Halichœrus grypus" by zoologists, whilst the Common Seal is known as "Phoca vitulina." The male of the former species grows to be as much as 10 feet in length, whilst that of the Common Seal rarely attains 5 feet. Both these seals breed on the British coast. The Common Seal frequents the north circumpolar region, being found on the northern coasts on both sides of the Atlantic, and also on both sides of the Pacific, and even makes its way down the coasts of France and Spain into the Mediterranean, where it is rare. A few years ago one appeared on the beach at Brighton! It may often be seen on the west coast of Scotland, of Ireland, Wales, and Cornwall, where it breeds in caves. Its hairy coat is silky, and has a yellowish-grey tint spotted with black and dark grey, most abundantly on the back.

The Great Grey Seal does not occur in the Pacific, but is limited to the northern shores on both sides of the Atlantic. Its coat is of a more uniform greyish-brown colour than that of the Common Seal, and when dried by exposure to the sun has a silvery-grey sheen. The Great Grey Seal is a good deal rarer on our coasts than is the Common Seal. It is now limited to the south, west, and north coasts of Ireland, to the great islands on the West of Scotland, the Orkneys, the Shetlands, and some spots on the east coast of Scotland. It is heard of as a rare visitor to the Lincolnshire "Wash," the coasts of Norfolk, Cornwall, and Wales. Some years ago (in 1883) I found a newly-born Grey Seal on the shore of Pentargon Cove, near Boscastle, North Cornwall. It appears that whilst (contrary to the statements of some writers) the Common Seal produces its young most usually in caves or rock-shelters, the Great Grey Seal chooses a remote sand island or deserted piece of open shore for its nursery. The Common Seal gives birth to its young—a single one or a pair—in June; the Great Grey Seal about the 1st of September. While the young in both species is clothed when born in a coat of long yellowish-white hair, this coat is shed in the case of the Common Seal within twenty-four hours of birth, exposing the short hair, forming a smooth, silky coat, as in the adult, and the young at once takes to the water and swims. On the other hand, the long yellowish-white coat of hair persists in the young of the Great Grey Seal for six or seven weeks, during which time it remains on shore, and refuses to enter the water. It is visited at sundown by the mother for the purpose of suckling it. According to Mr. Lyell, this renders the young of the Great Grey Seal peculiarly liable to attack by reckless destructive humanity, and he accordingly proposes legislation to render it a penal offence to destroy the young seals or the mothers during the nursing season. It is estimated that the total number of Great Grey Seals in Scottish waters has been reduced to less than 500, and that in English and Irish waters the total is even less.

It has often been desired by naturalists that a check should be put by the Legislature upon the wanton destruction of the common seal, as well as of the grey seal. It is certainly a regrettable result of the increased visitation of our remote rocky shores by holiday-makers, so-called "sportsmen" and thoughtless ruffians of all kinds, that the large, and perfectly harmless, grey seal is likely to be exterminated. In former times in these islands, as to-day in more northern regions, there was a regular "seal fishery," and vast numbers of seals were annually slaughtered for the sake of their skins and fat. The fur of both our native species, though differing vastly from the soft under-fur of the fur-seals, or Otariæ, of the North Pacific—which belong to a different section of the seal group, having small external "ears," and hind feet which can be moved forward and used in walking—is yet largely used for making gloves and thick overcoats.

To-day the number of British seals killed and brought to market is so small that no local fishery interests would suffer were all protected by the law during the spring and summer, when breeding and the rearing of the young is in progress. There is even less reason for objecting to the protection of the larger and rarer "Great Grey Seal," which, unless it had been placed under the shelter of an Act of Parliament, would in five or six years have ceased to be a denizen of the British Islands.

Owing to my having accidentally made the acquaintance of a young grey seal, as mentioned above, in North Cornwall, I feel a special interest in the legislative protection of this kind. I was at Boscastle at the end of August, and was delighted to see there on the morning after my arrival three or four of the common seal swimming in the little rock-bound harbour. I was told by native authorities that there was a cave in the rocks at the side of Pentargon Cove, a couple of miles distant (formerly inaccessible from the cliffs), where these seals breed, and that it had been the custom of some of the young men of the district to go round there in a boat when wind and tide served in the early spring and "raid" the cave. They could get in at low tide, and, armed with heavy cudgels, they would attack the seals which were congregated in the cavern to the number of thirty or forty. A single well-delivered blow on the nose was sufficient, I was assured, to kill a full-grown seal, and if fortunate the raiders might secure ten or a dozen seals, which were then sold for their skins and oil to Bristol dealers. The enterprise was dangerous on account of the rising tide and the struggles of the seals and their assailants among the slippery rocks and deep pools in the darkness of the cave. Cruel and savage as the adventure was, it yet had its justification on a commercial basis—similar to that claimed for other "fisheries" of the great beasts of the sea hunted by man for their oil and skins. The seals of this cave were undoubtedly the small common seal—the Phoca vitulina—and I gathered that little had been heard of late years of successful expeditions to these rocks. I was, however, told that a path had been cut and ropes fastened to iron stanchions in the face of the rocky cliffs of Pentargon Cove just before my visit to Boscastle, which rendered it now comparatively easy to descend the 150 feet of rock from the hill overlooking it and reach the shore of the curiously isolated and enclosed cove.

So, with two companions — my sisters—I set off the next morning for Pentargon Cove. We climbed down the face of the cliff by the aid of the much-needed ropes and found ourselves on the shore, the tide being low. We hoped that we should be able to get a view of the "seal-cave" and some of its inhabitants swimming in its neighbourhood. We were disappointed in this, and my companions hastened down to the water's edge, in order to get as near as possible to the rocky sides of the cove. I was about to follow them when I saw, lying in the open, on the pebbles above high-tide mark, what I took at first for a white fur cloak left there by some previous visitor. I walked up to it, when, to my extreme astonishment, it turned round and displayed to my incredulous gaze a pair of very large black eyes and a threatening array of teeth, from which a defiant hiss was aimed at me. It was a baby seal, covered all over with a splendid growth of white fur, three inches deep. He was twice as big as the fur-covered young of the common seal—more than two feet long—his black eyes were as big as pennies, and he was lying there on the upper beach, far from the water, in the full blaze of the sun, as dry and as "fluffy" as a well-dressed robe of Polar bear's skin. We were indeed well rewarded for our excursion in search of the seal's cave of Pentargon Cove! For this was a new-born pup of the Great Grey Seal, entirely unconnected with the inferior population of the inaccessible cave, laid here in the open by his mother at birth (as is the habit of her species), little suspecting that the long-secluded shore of Pentargon Cove had that year been rendered accessible to marauding land-beasts for the first time. Not knowing the peculiarities of the grey seal and the refusal of its young to enter the water until six weeks after birth, when it sheds its coat of long white hair, we cautiously rolled the little seal on to my outspread coat and carried him to the water's edge. After the hissing with which he had greeted my first approach he was not unfriendly or alarmed, and for my part I must say that I have never yet stumbled upon any free gift of Nature which excited my admiration and regard in an equal degree. His eyes were beautiful beyond compare. We placed him close to the water and expected him to wriggle into it and swim off, but, on the contrary, he wriggled in the opposite direction, and slowly made his way, by successive heaves, up the beach. He was not more than a day or two old, as was shown by the unshrunken condition of the umbilical cord. We did not like to leave him exposed to the attacks of vagrant boys, who might climb down into the cove, so we carried him on my coat to the shelter of some large rocks, a hundred yards along the shore. There, with much regret, we left him.

But on the following evening, as we sat down to dinner, I heard from some other visitors at the Wellington

Inn, to whom, under pledge of secrecy, I had confided our discovery, that they had been to Pentargon Cove to visit our young friend, and found that he had been removed (probably by his mother) back to the exact spot where we had found him. They also stated that his presence there had become known in the village, and that the conviction had been expressed that "the boys" would certainly go and stone him to death! I had already reproached myself for going elsewhere that day instead of to Pentargon Cove to look after my young seal, and now I hastily left my dinner, procured in the village two men and a potato sack, and hurried to Pentargon Cove. As we approached the edge of the cliff the sun was setting, and the cove was very still and suffused with a red glow. Then a weird sound rent the air, like that made by one in the agonies of sea-sickness. It was the little seal calling for his mother! It is the habit of the females of this species to leave the shore during the day when they go in search of the fish on which they feed, and to return to their young in the evening, in order to suckle them. I could see, from above, my baby friend—a little white figure all alone in the deepening gloom of the great cliffs—raising his head and, by his cries, helplessly inviting his enemies to come and destroy him. In a few minutes we were down by his side, had placed him in the potato sack, and brought him to the upper air. On the way to the inn I purchased a large-sized baby's bottle with a fine indiarubber teat. We placed the little seal on straw in a large open packing-case in the stables, whilst the kitchen-maid warmed some milk and filled the feeding-bottle. Then I brought it to him, looking down on his broad, white-furred head, with its wonderful eyes, set so as to throw their appealing gaze upwards. I touched his nose with the milky indiarubber teat. With unerring precision his lips closed on it, his nostrils opened and shut in quick succession, and he had emptied the bottle. I gave him a quart of milk before leaving him and getting my own belated meal. He slept comfortably, but at four in the morning his cries rent the air, and threatened to wake every one in the hotel. I had to get up, descend to the kitchen, warm some more milk for him, and satisfy his hunger. He became fond of the bottle, and also of the friend who held it for him. I arranged to take him to the Zoological Gardens when, after three days, I left Boscastle. He travelled to London in the guard's van in a specially constructed cage, and was as beautiful and happy as ever when I handed him over to the superintendent at Regent's Park.

In those days (as it happened) there was little understanding or care at "the Gardens" as to the feeding of an exceptional young animal like my little seal. It is possible to treat cow's milk so as to render it suitable to a young carnivore, much as it is "humanized" for the feeding of human babies, and I was willing to pay for a canine foster-mother were such procurable. I had then to leave London in order to preside over one of the sections of the British Association's meeting at Southport, and intended to take complete charge of my baby seal upon my return. But in less than a week the neglectful guardians at Regent's Park had killed him with stale cow's milk. I believe such a foundling would have a better chance there to-day, but the rearing of young mammals away from their mother is, of course, a difficult and uncertain job.

I do not regret having taken the baby seal from Pentargon Cove, for I undoubtedly saved him from a violent death, whilst his mother would soon recover from the loss due to my action—a loss to which she and her fellow "grey seal-mothers" must be not unfrequently exposed from other causes. I do regret, however, that it did not occur to me until too late that it would have been a wonderful experience to lie quietly on the shore some few yards from the baby seal, as the sun set, and then to see and hear the great seal-mother—7 or 8 feet long—swim into the cove, raise her gigantic bulk on the shore, and heave herself across the pebbles to her eager child. To witness the embraces, caresses, and endearments of the great mysterious beast would have been a revelation such as a naturalist values beyond measure. And so I hope, with all my heart, that Mr. Lyell will succeed in his good work of protecting the Great Grey Seal.

CHAPTER V

THE GROUSE AND OTHER BIRDS

IN August when so many people are either shooting or eating that delectable bird—the grouse—a few words about him and his kind will be seasonable. "Grouse" is an English word (said to have meant in its original form "speckled"), and by "the" grouse we mean the British red grouse, which, though closely related to the willow grouse, called "rype" (pronounced "reepa") in Norway—a name applied also to the ptarmigan—is one of the very few species of birds peculiar to the British Islands. The willow-grouse turns white in winter, and is often called the ptarmigan, which it is not, though closely related to it. The willow-grouse inhabits a sub-arctic zone, which extends from Norway across the whole continent of Europe and Asia, and through North America, from the Aleutian Islands to Newfoundland. The red grouse does not naturally occur beyond the limits of the British Islands. It does not turn white in winter, and the back of the cock bird is darker in colour, as is also the whole plumage of the hen bird, than in the willow-grouse. The red grouse lives on heather-grown moors; the willow-grouse prefers the shrubby growths of berry-bearing plants interspersed with willows, whence its name. No distinction can be discovered in the voice, eggs, build, and anatomical details of the two species. The red grouse and the willow-grouse were, at no very distant prehistoric period, one species, but the race which has become isolated in these islands has just the small number of marked differences which I have mentioned, and it breeds true, and therefore we call it a distinct "species." In Scotland, the red grouse is called "muir-fowl," and a century ago was almost invariably spoken of in England as moor-fowl, or moor-game. It is found on moors from Monmouthshire northward to the Orkneys, and inhabits similar situations in Wales and Ireland.

The red grouse and the willow-grouse belong to a section or "order" of birds which are classified together because they all have many points in common with "the common fowl" or jungle-cock and the pheasants. That order or pedigree-branch was named by Huxley Alectoromorphæ, or cock-like birds, perhaps more simply termed Galliformes, Gallus being the Latin name for "chanticleer." When there is a question of the groups recognized in the classification of animals, it is well to bear in mind, once for all, that the biggest branches of the animal pedigree are called "phyla" (or sub-kingdoms); that these have branches or sub-divisions which are called "classes" (birds are a class of the phylum Vertebrata). Classes divide into "orders" these often are subdivided into "sub-orders." Orders comprise each several smaller branches called "families,' families branch into "genera," and each "genus" contains a number of "species" which have diverged from a common ancestral form, and become more or less stable and unchanging (but not unchangeable) at the present day. The individuals of a species are distinguishable by certain marks, shape, and colour from the individuals of other species of the genus. They breed true to those points when in natural conditions, and show some differences of habit, locality, and constitution which emphasize their distinction as a separate "species."

The order Galliformes of the class Aves or birds is one of some eighteen similar orders of birds. It contains several families, namely, the grouse-birds, the partridges, the francolins (formerly introduced into Italy from Cyprus), the quails, the pheasants, including the common fowl or Gallus, the peacocks, the turkeys, and, lastly, the guinea-fowls. The mound-builders and the South American curassows (very handsome birds to be seen at the Zoological Gardens) are families which have to be separated from the rest as a distinct sub-order. Fifty years ago the pigeons were placed in one order with the galliform birds, which was termed "Rasores," or scratching birds; but they are now separated under the name Columbiformes.

All the galliform birds are specially agreeable to man as food, and the domesticated race of the jungle-fowl—for which we have no proper English name, except that of "the" fowl[1]—is second only to the dog in its close association with man. It seems to have been domesticated first in Burma, and was introduced into China about 1000 B.C., and through Greece into Europe about 600 B.C. It is not mentioned in the Hebrew Scriptures, nor by Homer, nor figured on ancient Egyptian monuments. It was called "the Persian bird" by the Greeks, indicating that it came to them from the Far East through Persia. The common or barn-door fowl is assigned to the genus Gallus, of which there are four wild species. It is very closely related to the pheasants (genus Phasianus, with several "local" species); indeed,

[1] "Chanticleer" is the name given to the cock-bird of this species in the very ancient story of "Renard the Fox."

so closely that, when pheasants and "fowls" are kept together in confinement they will sometimes interbreed and produce vigorous hybrids. The peacocks are Indian, and with them is associated the Malay Argus-pheasant. They share with the turkeys, which are North American in origin, the habit of "display" by the male birds when "courting"—a habit which we see in a less marked form in the strutting, wing-scraping, and cries of the pheasants, chanticleers, and grouse-birds. The various species of partridges are confined to the temperate regions of the Old World, but the word is wrongly applied in America and Australia to other kinds of birds. The guinea-fowls are African, and so are the francolins and quails, the latter migrating to the South of Europe. It is an interesting fact that, when the turkey was first brought from America, about 1550, a confusion grew up in Europe between it and the guinea-fowl. The turkey was given a genus (Meleagris) to itself by Linnæus, who called it "M. gallopavo," whilst the guinea-fowl was called "Numida meleagris." We know, at present, other "species" of Meleagris besides M. gallopavo, and other species of Numida.

Now we revert to the grouse-birds, a family for which the zoologist's name is Tetraonidæ. They all have the beautiful crimson arch of bare knobby skin above each eye which gives its chief beauty to our grouse. The family contains several genera and included species. The largest species is the capercailzie (a Gaelic word), or cock of the wood, called by the French "coque du bois," by the Germans "auerhahn" (auerhuhn for the hen bird), and by the Norwegians "tiur." It is placed in the genus Tetrao (which gives its name to the "family"), and receives the specific name "urogallus." This fine bird was formerly native in England, as well as in Scotland and Ireland, and is found in the pine forests of Europe from Spain to Lapland and Greece. It has been re-established in Scotland since 1838. An allied species is found in Siberia. The black grouse (often called black cock and grey hen) is a second species of the genus Tetrao, namely, T. tectrix. It is often called "Lyrurus tetrix." The French name for it is "coq de bruyère"; the German is "birkhahn." It is a smaller bird than the capercailzie, but frequently produces hybrids with that species. The beautifully curled tail-feathers are favourite adornments for the hat of mountaineers and hunters in the Tyrol and Switzerland.

Though the word "grouse" may have been first applied (as some think) to the black cock, it is now the proper appellation of the red grouse. This bird is placed by zoologists in the genus Lagopus—the members of which are easily distinguishable from other Tetraonidæ by the fact that their feet and toes are well covered with feathers. "L. scoticus" is the scientific name of the red grouse. Being a purely British bird, it has no foreign designations. "L. saliceti" is the name of the allied willow-grouse, which has an endless variety of names, owing to its great range of distribution. The willow-grouse is often called ptarmigan, and is sold as such to the number of thousands by poulterers in our markets, but it is not the true ptarmigan. Owing to the fact that its plumage is quite white in winter, there is much excuse for the confusion. The name "ptarmigan" is the Gaelic word "tarmachan," and no one has explained how the initial "p" came to be added to it. The bird called in Scotland tarmachan or ptarmigan is a third species of Lagopus. It is much rarer in Scotland than the red grouse, and lives in high, bare ground. It is numerous at an elevation far above the growth of trees in Norway, and occurs also in the Pyrenees and the Alps. It turns white in winter (as do all the species of Lagopus except the red grouse), and differs in many features of structure from the red grouse and the willow-grouse. It is called "L. mutus." A fourth species of Lagopus is L. rupestris, of North America, Greenland, Iceland, and Siberia. Spitzbergen has a fifth species, L. hemileucurus, a large form. The sixth and smallest species of Lagopus is the L. leucurus of the Rocky Mountains. There are yet further some excellent grouse-like birds, which are separated to form other genera distinct from Lagopus. Though they do not inhabit the British Islands, some of them are brought occasionally to the London market. The hazel-hen of continental Europe is one of these, and is considered to be the most delicate game-bird that comes to table. It is placed in the genus Bonasa, and receives the specific name "sylvestris." The French call it "gelinotte" (under which name various kinds of cold-storage grouse are often served in London clubs and restaurants), the Germans "hasel-huhn," and the Scandinavians "hjerpe." It is a purely forest bird. It is represented in North America by four other species, of which the best known is Bonasa umbellus, called by the Americans the ruffed grouse or birch-partridge.

Another genus of Tetraonidæ, or grouse-birds, is called "Canachites," and contains the species known as the Canadian spruce-partridge, Franklin's spruce-partridge, and the Siberian spruce-partridge. Nearly allied to these is a genus Dendragapus, with three North American species. Then we have the sage-cock of the plains of California (Centrocerus urophasianus), three species of sharp-tailed grouse (genus Pediocætes), and "the prairie hen," of which three species are placed in the genus Tympanuchus. The United States have, undoubtedly, a great variety of grouse-like birds. Nevertheless, a year ago I met in Paris an American from the neighbourhood of Boston who told me that he should have to desert his native land and come to live in Europe, because he could not obtain a regular supply of game-birds for his table in the eastern States. He was eating a Scotch grouse at the time with evident satisfaction.

The supply of grouse in this country has been threatened by disease caused by the attempt to make the moors carry more birds than they would do under natural conditions. The number annually shot on British moors is enormous. Predaceous animals have been destroyed in order to increase the number of birds, but this proceeding has resulted in allowing the weakly to survive. The undisturbed stretches of moorland have also of late years been greatly broken into both by roads and building, and by the too abundant visitation of strangers of all kinds. Only a few years ago one moor-owner was able to boast that he had on several occasions killed over 500 head of grouse in a single day on his moor, and that in one season he and his guests had killed 18,231 head of grouse on that same moor! Personally I rejoice when grouse are abundant, but it seems to me possible that the moor above mentioned had been made to carry, so to speak, too heavy a crop. However, there is reason to hope that the balance of Nature is restored after a few years of disease, which kills off the too-abundant bird population.

CHAPTER VI

THE SAND AND PEBBLES OF THE SEASHORE

THE "beach" on our English coast is an accumulation of pebbles or of sand, or of both, often accompanied by dead shells and other fragments thrown up by the sea. Very generally it slopes rapidly from above high-water mark to about half-tide limit, and then merges into a more horizontal expanse of fine, compact sand. This last is not "a beach" thrown up by waves, but a sediment or deposit. It forms a flat, often ripple-marked plain (much has been written as to how those ripple-marks are produced), which is exposed at low water, the sea retreating for a quarter or even half a mile or more over it, on some level shores. Sometimes, though rarely, the sea rises and falls against a hard, rocky cliff without forming any beach or exposing any "shore" even at low tide. This occurs on parts of the Cornish coast, where the Atlantic beats against adamantine cliffs, which even at low tide rise sheer from the water. Again, it sometimes happens that the shore is simply formed of a terrace of sloping hard rock, without any "beach." But on the coast of England generally there is a good beach of sand or pebbles, or both, overlying the native rock or clay, and sometimes it is growing every year, so as to extend the land surface seawards and add new acres to the possessions of the landlord.

On other parts of the coast the beach "travels," being driven along the underlying solid shore by the prevailing direction of the tidal currents and by the waves. The sea-waves break close to the soft cliffs of clay, sand, and sandstone. These are continually crumbling away owing to the action of land water, which soaks from the surface down to the layers of clay and forms subterranean springs and streams. They undermine the face of the cliff and cause the upper parts to topple. When there is a big, broad, growing beach in front of such a cliff, the breaking down or "toppling" of its face only leads to the formation of a slope (at the "angle of rest"), and things remain but little changed for ages. But if the beach is not being piled up and added to and growing out seawards year by year, and is, on the contrary, a travelling beach, then the sea comes close up to the cliff, and when masses of it topple on to the beach the sea washes them away, and no "slope of repose" is formed. The cliff keeps on toppling as it is undermined by springs of land water. Its natural buttress against further breakage — namely, its own fallen material—instead of resting against it as a great sloping, protective bank, is washed away by the sea as fast as it falls, and is carried down the coast by the tidal currents. This is the story of "coast erosion" about which there has recently been a Government inquiry. Where the combined action of prevailing winds and sea currents is throwing up and adding to the beach there is no coast erosion. The causes of the sea currents on our coasts are not easy to determine, as they are connected with the general contour of the land and the currents in large tracts of sea, such as the Channel and the North Sea. Coast erosion is a serious thing. Large parts of the coast of Suffolk and Norfolk are being thus washed away. It can be prevented by "holding" the beach with piles and boarding, but this costs too much to make it worth doing unless the land so preserved has a special value for the erection of houses.

At Felixstowe, where I am writing, the sea has swept away most of the flat—the "dunes," or "deans"—covered with grass, which it had itself built up by a contrary accumulating action before the time of the Romans. On this flat the ancient Roman town was built. Why the sea has reversed its action is very difficult to say. But within my knowledge of this place high-water mark has advanced as much as 300 yards nearer than it was to the old roadway and to old houses. The great town of Dunwich, which in the Middle Ages had eleven churches, strong fortifications, and a flourishing trade, stood on the flat grass-land in front of the cliff on the Suffolk coast. Its site is now under the sea, not far from here. The breaking away of the cliff (on to which part of the town extended) is still going on there. A few years ago I saw a great bricked well lying like a fallen chimney on the shore. It had been exposed by the crumbling of the cliff, and at last fell out of it. Once that well supplied fresh water to the monastery, part of the walls of which are still standing, and were formerly three-quarters of a mile distant from the sea-shore. The prehistoric cliffs to which the sea came before it formed the flats or links which it is now again eating away, are often traceable a mile or two inland. On the other hand, on parts of the Lincolnshire coast the sea has piled up sand and shingle and added valuable land to the extent of hundreds of acres to the property of those whose estates were bounded by the shore line, and is still doing so. Perhaps the action of the north wind in blowing back and piling up sand out of the reach of the tide is influential in producing this increase of shore-lands, which face northwards. Blown sand forms hills 30 feet and more in height on such flat lands as those of the Sandwich and Deal "links," which have been thrown up by the sea since St. Augustine landed at Richborough, then a seaport, now a couple of miles from the sea. On the French coast near Boulogne the sand has been blown inland so as to form stratified deposits on the low hill country as far as 3 or 4 miles from the sea, and the neighbouring port of Ambleteuse, which five hundred years ago had the chief trade with England—is now nothing but a vast stratified "dune" of blown sand. The great Napoleon made some attempt to reopen the harbour, but gave it up as a bad job; the blowing of sand inwards from the enormous tract of flat, sandy shore was too much for his engineers.

The "erosion" and the contrary process of the "extension" of the coast by the action of the waves and currents of the sea must be kept apart and distinguished from a process leading to similar but not identical results, namely, the actual "crumpling" or "buckling" of the earth's crust, leading to the rising of the land surface in some parts of the globe relatively to the sea-level, and on the other hand to the sinking of the land beneath the sea in other regions. This change of the actual level of the land has continually gone on in the past, and is continually going on to-day. What are called "raised beaches" are seen on many parts of the coast. These are lines of ancient beach, consisting of sea-worn pebbles, fragments of shell, etc., forming terraces along the face of the rocks which rise from the present seashore—terraces which are now 15, 30, or more feet above the sea-level, although they must at no very distant period have been at the level of the sea. The land has risen and carried them up out of reach of the waves. Such a raised beach is seen along the rocks bordering Plymouth Sound, at a height of some 15 feet (so far as I can, at this moment, remember) above high-water mark. Owing to the fact that the rock is limestone, and is dissolved and redeposited by rain water, as a rock of sugar might be, the pebbles and shells of the old beach are all stuck together or "petrified" by re-deposited limestone (carbonate of lime). Lumps of it can be carried away as specimens.

Geological deposits of much older date than these comparatively recent raised beaches tell us of the rising of great masses of land. Thus, for instance, marine shells in a deposit not quite so old as our chalk cliffs and downs, are present at a height of 10,000 feet, forming part of the Alps. At one time that very spot was the bottom of the ocean, whilst other tracts of the earth's surface, now sunk hundreds of fathoms below the sea-level, stood out as continents, with hills and valleys well raised above the waters. Direct evidence of the recent sinking of the coast as distinct from its erosion is not familiar to us in England. The evidence of it is naturally obliterated, as the sinking goes on, whereas on a rising coast the evidence is as naturally preserved. But on the shores of the Mediterranean near Naples the evidence of sinking is well preserved, and has been carefully studied and recorded. The ancient Roman road is still sunk beneath the water, though the celebrated temple of Puteoli, which was formerly submerged by the sinking of the land, has reappeared by a subsequent elevation of the same area. This has not brought the site to so high a level as it had when the temple was built, as appears from the fact that the Roman paved roadway close by is still some 15 feet below the surface of the sea.

A beach is built up of water-worn pebbles, consisting usually of bits of the rock of the immediate vicinity, which have become rounded and shaped by continually rolling and knocking against one another as the waves of the sea throw them up or drag them down the sloping heap of like pebbles which is accumulated near high-water line. At Dover and such places, under chalk cliffs, the beach consists of chalk pebbles oval in shape, often of 8 or 9 inches in length, with a large number of well-rounded flint pebbles as big as your fist interspersed, or outnumbering the chalk pebbles. At Tenby, in South Wales, the beach consists of assorted sizes of limestone pebbles, well-worn bits of the limestone cliffs of the neighbourhood. Large numbers of them are literally "worm-eaten," being bored into, hard and dense as they are, by a little marine worm (known as Polydora), which may be sometimes found alive and at work in these limestone pebbles lying between tide limits, or more easily at other places in similarly placed chalk blocks or pebbles. On a coast bounded by granite cliffs you get a beach of granite pebbles; where there are cliffs of slate or of sandstone, pebbles of slate or of sandstone.

But there are some beaches which, as remarked above, are continually travelling along the coast. That on the English shores of the North Sea, for instance, is always moving southwards, except where it is held by piles and breakwaters, locally called "shies." Moreover, the land of the East Coast, especially the Suffolk and Norfolk coast, in the course of its erosion, has given back to the sea old deposits of the glacial and post-glacial period, consisting of gravels and "drift," made up of flint pebbles and fragments of rocks from the more northern regions over which the great European ice-cap of the glacial epoch extended, and from which it ground and tore the surface rock and carried large and small masses—boulders and incredible millions of tons of broken up fragments—and spread them over East Anglia (where they form the so-called "glacial drift"), and over regions still submerged in the North Sea. Consequently the beach on the Suffolk seashore has a specially variegated assortment of pebbles from all sorts of more northerly situated rocks—though small flint pebbles, derived directly from glacial drift and by the drift from the chalk land-surface (the chalk itself not now reaching the shore-line of East Anglia), are greatly predominant. It is in the chalk that flint takes its origin, being found there as large irregular nodules and sheets.

CHAPTER VII

THE CONSTITUENTS OF A SEABEACH

I ONCE went down to Aldeburgh, on the Suffolk coast, with a party of friends, which included an American writer, himself as delightful and charming as his stories. Why should I not give his name? It was Cable, the author of "Old Creole Days." We walked through the little town to the sea-front, and came upon the immense beach spreading out for miles towards Orford Ness. "Well, I never!" said he to me; "I suppose the hotel people have put those stones there to make a promenade for the visitors. It's a big thing." It took me some time to persuade him that they were brought there by the sea and spread out by it alone. It was his first visit to Europe, but he had seen the seashore on the other side, and there was nothing like this over there, he declared. A similar readiness to ascribe Nature's handiwork to the enterprise of hotel-keepers led a visitor to the Bel Alp, in the Rhone Valley, when he looked down from that high-placed hostelry on to the great Aletsch glacier, with its central "moraine" of huge rock masses and debris, to exclaim, "I see the proprietor has spread a cinder-path along the glacier to prevent us from slipping. It's a convenience, no doubt, but gives a nasty dirty look to the snow." Mr. Cable, when he once realized that the great Aldeburgh beach was a natural production, did what a true poet and naturalist must do—he fell in love with it, and spent hours in filling his pockets with strange-looking pebbles of all kinds until he was brought into the house to dinner by main force, when he spread his collection on the table, and demanded an explanation of "what, whence, and why" in regard to each pebble. Our companions—a great lawyer, a military hero, a politician, and two "learned men"—regarded him as eccentric, not to say childish. But I entirely sympathized with him, and when next day we sailed down to Orford and stood in front of the old Norman fortress, he further established himself in my regard by deeply sighing and exclaiming, "So that is a real English castle!" whilst several large tears quietly streamed down his undisturbed countenance.

To give an idea of what various rocks from far-distant localities may be brought together on an East Coast beach, take that of Felixstowe as an example. What is true of the East Coast is to some extent also true of the South Coast, and, indeed, wherever the sea makes the pebbles of a modern beach from the materials furnished by the breaking up of old deposits, which were in their day brought by ice-flows or torrential currents from remote regions. The most abundant kind of pebbles on the Felixstowe beach are small, rounded, somewhat flat pieces of flint, derived not directly from the chalk which is the "stratum" or "bed" in which flint is originally formed, but from the Red Crag capping the clay cliffs (London clay or early Eocene), and also from surface washings and "gravels" (of later age than the crag) farther north, whence they have travelled southward with many other constituents of the beach. All these flints are stained ruddy brown or yellow by iron—a process they underwent when lying in the gravels or in the crag in which they were deposited as pebbles, broken, washed, and rolled ages ago from the chalk. The iron is in a high state of oxidation, and stains not only flint pebbles but the sands of the Red Crag and later gravels a bright orange-red, or sometimes a less ruddy yellow. The iron comes originally from very ancient igneous rocks in which it is black and usually combined with silica. The chalk flints are always, owing, it seems, to minute quantities of carbon, quite black in the mass, but thin, translucent splinters have a yellowish-brown tint. The flints are free from iron stain when taken direct from the chalk. The commonest pebble next to flint is milky quartz, or opaque white quartz. This is derived from some far northern source, where there are igneous rocks traversed by veins of this substance (perhaps Norway). Quartz, like flint, is pure silica, the oxide of the element silicon. It appears in another form as rock-crystal, and also as chalcedony and agate. Opal also is pure silica, but differs from quartz and its varieties in being non-crystalline or amorphous, and in being less hard and of less specific gravity than quartz. Opal is soluble in alkaline water containing free carbonic acid, such as are many natural waters and the sea! But quartz is not so. The siliceous "spicules" and skeletons of many microscopic animals and plants are "opal." The gem known as "opal" is a variety owing its beauty to minute fissures in its substance which break up light into the prismatic colours.

A great deal rarer than the milky quartz, but well known on the East Coast on account of their beauty, and often sought for to be cut and polished, are the small rolled bits or pebbles of chalcedony or agate, which have been bedded before their appearance on the beach in some of the pre-glacial or post-glacial gravels, together with the flints, and in consequence are often stained of a fine red. Such clear red-stained chalcedony is called "carnelian"; if the banded agate structure shows, it is called agate rather than carnelian. It is wonderful how many beautiful pieces of both carnelian and agate are picked up on the Felixstowe beach, rarely, however, bigger than a hazel nut. The original source of these carnelians and agates is the East of Scotland. At Montrose you may see the igneous rock containing pale, lavender-coloured agate nodules as big as a potato, the breaking and rolling of which by the sea into small bits has furnished our Suffolk carnelians. Quartzite—more or less translucent, sandy-looking pebbles, colourless or yellow: jasper, black or green with red veining: a fine wine-red or purple stone often veined with quartz—are all more or less common, and come from northern igneous rocks—possibly some from Scandinavia and some from the breaking up of an ancient "breccia" of the Triassic age, which still exists northwards of East Anglia.

Other pebbles very common on this shore are those formed in a curious way by the sea-water from the clay cliffs and sea bottom which are here present, and are of that special geologic age and character known as the London clay. The sea at this moment is continually converting the clay of our Suffolk shore into "cement-stone" by a definite chemical process. The clay and many other things submerged in the sea, as Shakespeare knew, "undergo a sea-change." The cement-stone used to be dredged up from the sea bottom and ground to make cement at Harwich. Great rock-like slabs of it pave the shore at low water, and pebbles of it are abundant. The curious thing is that ages ago—geological ages, I mean—when the sea was throwing up here the old shell-banks and sand-banks known nowadays as "the Red and Coralline Crags," the London clay cliffs and clay sea bottom were in existence just as they are now. But in that period there existed here enormous quantities of bones of whales of kinds now extinct, which had lived a little earlier in the sea of this area, and were deposited in vast quantity as a sort of first layer of beach or shallow water sea-drift. Bones consist largely of phosphate of lime, and are used as manure. In that old crag sea the phosphate of lime was dissolved from the deposit of bones, and as we find occurring in the case of other clays and other bones elsewhere—was chemically taken up by the clay—the same kind of clay which to-day is being converted into "cement-stone." It was thus, at that remote period, converted into "clay phosphorite," owing to the presence of the immense deposit of whales' bones, and it has been known for sixty years as Suffolk "coprolite," owing to a mistaken notion that it was the petrified dung of extinct animals. It has been dug up by the ton from below the crag all over this part of Suffolk, where it forms, together with bones, teeth, flints, and box-stones, a bed of small nodules, a foot or so thick separating the London clay from the shelly "crag." This bed is called the Suffolk bone-bed or nodule-bed. The phosphorite, or "coprolite," occurs in the form of bits of clay, hardened by phosphate of lime, and of the colour of chocolate, and hundreds of tons of it have been used by manufacturers of the manure known as "superphosphate." Henslow, of Cambridge, Darwin's friend and teacher, was the first to point out its value. Bits of it, as well as box-stones, and fragments of bone, teeth of whales, of sharks, of mastodon, rhinoceros, tapir, and other extinct animals—all fallen from the bone-bed in the cliff—are found mixed with the pebbles of the Suffolk beach by those who lie on that beach in the sunshine, and, for want of something better to do, turn over handful after handful of its varied material. And, besides all the stones I have already mentioned, they find amber, washed here by some mysterious currents from the Baltic, wonderful fossil shells out of the crag, the cameo shell, and the great volute, — shells which are as friable as the best pastry when dug out of the Red Crag, but here on the shore become hardened by definite chemical action of the sea-water, so as to be as firm as steel. Here, too, the "chiffonier" of the sea-shore finds recent shells, recent bones (slowly dissolving and wearing away), well-rounded bits of glass, jet drifted down from Whitby, Roman coins, bits of Samian ware (!), mediaeval keys, bits of coal, burnt flints (from steamers' furnaces), and box-stones.

A very important and interesting thing about "beaches" is the way in which the pebbles of which they consist are assorted in sizes. Suppose that one prepares a trough some two or three yards long and twelve inches deep, and lets it fill with water from a constantly running tap, tilting it slightly so that the water will overflow and run away at the end farthest from the tap. Then if one drops into the trough near the tap handful after handful of coarse sand and small stones of varied sizes, they will be carried along by the stream, and the more rapid and voluminous the stream the farther they will be carried. But they will eventually sink to the bottom of the trough, the bigger pieces first, then the medium-sized, then the small, and the smaller in order, as the current carries them along, so that one gets a separation and sorting of the solid particles according to size, a very fine sediment being deposited last of all at the far end of the trough. The waves of the sea are continually stirring up and assorting the constituents of the beach in this way. Usually the largest pebbles are thrown up farthest by the advancing waves, and dropped soonest by the backward suck of the retreating water, so that one generally finds a predominance of big pebbles at the top of the beach. But on the flat shore of firm ripple-marked sand lying lower down than the sloping "beach" and only exposed at quite "low tide," one often finds very big pebbles of eight or nine pounds weight scattered here and there and little rubbed or rounded. They have gradually moved down the sloping beach and are too heavy to be thrown back again by the waves of the shallow sea which flows over the flat shores characteristic of much of our south-eastern and southern coast. On some parts of the coast huge banks, consisting exclusively of enormous pebbles as big as a quartern loaf, are piled up by the waves, forming a great ridge often miles in length, as at the celebrated Chesil pebble bank near Weymouth, and at Westward Ho! in North Devon. The presence of these specially large pebbles is due to the special character of the rocks which are broken up by the sea to form them, and to the specially powerful wave-compelling winds and tidal currents at the parts of the coast where they are produced.

One generally finds a selected accumulation of moderate-sized pebbles lower down the beach as the tide recedes, and then still lower down patches of sand alternating with patches or tracts of quite small pebbles not much bigger than a dried pea. They are always assorted in sizes, but the extent of each tract of a given size of pebble varies greatly on different beaches along the coast, and even from day to day on the same shore. The greater or less violence of the waves, and of the currents caused by wind and tide, is the cause of this variation and local difference. The pebbles of the "beach" are, of course, always being worn away, rounded and rubbed down by their daily movement upon one another, caused by the waves as the tide mounts and again descends over the shore. Even the biggest stones, excepting those which lie in deeper water beyond the beach, are eventually rubbed down, and become quite small; but a point is reached when, the weight of the pebbles being very small, they have but little effect in rubbing down each other, and consequently where the pebbles consist of very hard material—like flints—the smallest ones are not so much rounded, but are angular and irregular in shape.

Whilst a perfect gradation in size can be found from the largest flint pebbles some 6 inches or 7 inches long to the smallest, usually not bigger than a split pea (though sometimes a patch of even smaller constituents may be found), there is a real break or gap between "pebbles" and "sand." I am referring now to what is commonly known as "sand" on the southern part of the East Coast, much of the South Coast, and the shores of Holland, Belgium, and France. There are "sands" of softer material (limestone and coral sand), but the sands in question are almost entirely siliceous, made up of tiny fragments of flint, of quartz, agate, and hard, igneous rock. They are often called "sharp" sand. The particles forming this sand are sorted out by the action of moving water, and form large tracts between tide-marks looking like brown sugar, for which baby visitors have been known to mistake them, and accordingly to swallow small handfuls. The strong wind from the sea blows the sand thus exposed, as it dries, inland out of reach of the tide, to form sand-dunes, and it is also deposited, together with still finer particles (those called "mud"), on the shallower parts of the sea bottom. The curious thing about the particles of "sharp" sand is that they are angular, and for the most part without rounded edges. If you examine them under a microscope you will see that they do not look like pebbles—in fact, they are not pebbles, for they are so small and have so little weight, or, rather, mass, that they do not rub each other to any effect when moved about in water. They look like, and, in fact, are, for the most part broken bits of silica, unworn and sharp-edged splinters and chips, glass-like in their transparency and most of them colourless, a few only iron-stained and yellow. Amongst these are a few rounded, almost spherical pieces, which are no doubt of the nature of minute water-worn pebbles. Although these few minute pebbles exist among the sharp, chiplike particles of "sand," it is clear that we must broadly distinguish "pebbles" of all sizes down to the smallest—from the much smaller "sand particles." There is no intermediate quality of material between "sand" and the finest "shingle."

CHAPTER VIII

QUICKSANDS AND FIRE-STONES

THERE are curious facts about sand which can be studied on the seashore. There are the "quick-sands," mixtures of sand and water, which sometimes engulf pedestrians and horsemen at low tide, not only at the Mont St. Michel, on the Normandy coast, but at many spots on the English, Welsh, and Scotch coasts. Small and harmless quicksands are often formed where the sand is not firmly "bedded" by the receding sea, and the sea-water does not drain off, but forms a sort of sand-bog. Then one may also study the polishing and eroding effect of dry blown sand, which gives a "sand-glaze" to flints, and in "sand-deserts" often wears away great rocks. The natural polishing of flints and other hard bodies by fine sand carried over them for months and years in succession by a stream of water, is also a matter of great interest, about which archæologists want further information.

A very interesting fact about the ordinary sand of the seashore is that two pints of dry sand and half a pint of water when mixed do not make two pints and a half, but less than that quantity. If you fill a child's pail with dry sand from above the tide-mark, and then pour on to it some water, the mass of sand actually shrinks. The reason is that when the sand is dry there is air between its particles, but when the sand-particles are wetted they adhere closely to each other; the air is driven out, and the water does not exactly take an equivalent space, but occupies less room than the air did, owing to the close clinging together of the wet particles. If you add a little water to some dry sand under the microscope, you will see the sand-particles move and cling closely to one another. "Capillary attraction"—the ascent of liquid in very fine tubes or spaces—is a result of the same sort of adhesive action. If you walk on the firm, damp sand exposed at low tide on many parts of the seashore when it is just free from water on the surface, you will see that when you put your foot down the sand becomes suddenly pale for some seven inches or so all round your foot. The reason is that the water has left the pale-looking sand (dry sand looks paler than wet sand), and has gone into the sand under your foot, which is being squeezed by your weight. The water passing into that squeezed sand enables its particles to sit tighter or closer together, and so to yield to the pressure caused by your weight. You actually squeeze water "into" the sand, instead of squeezing water "out" of it, as is usually the case when you squeeze part of a wet substance—say a cloth or a sponge. When you lift your foot up, you find that your footmark is covered with water—the water you had drawn to that particular spot by squeezing it. It separates as soon as the pressure is removed.

Quartz and quartzite pebbles occur on the South as well as the East Coast. They are sometimes called "fire-stones," because they can be made to produce flashes of flame. If you take a couple of these pebbles, each about as big as the bowl of a dessert-spoon (a couple of flint peebles will serve, but not so well), and holding one in each hand in a dark room, or at night, scrape one with the other very firmly, you will produce a flash of light of an orange or reddish colour. And at the same time you will notice a very pecular smell, rather agreeable than otherwise, like that of burning vegetable matter. It would seem that the rubbing together of the stones produces a fine powder of some of the siliceous substance of the stone and at the same time a very high temperature, which sets the powder aflame. I had the idea at one time, based on the curious smell given out by the flashing pebbles, that perhaps it was a thin coating of vegetable or other organic matter derived from the sea-water which burns when the stones are thus rubbed together; but I found on chemically cleaning my pebbles, first with strong acid and then with alkali, that the flame and the smell were produced just as well by these chemically clean stones as by those taken from the beach. The flame produced by the rubbing of the two stones seemed then to be like the sparks obtained by strike-a-lights of flint and steel, or the prehistoric flint and pyrites. Now, however, a new fact demands consideration. The supposition that the powdered silica formed, when one rubs the two pebbles together, is actually "burnt," that is to say, combined with the oxygen of the air by the great heat of the friction, is rendered unlikely by the fact that if you perform the rubbing operation in a basin of water with the stones submerged, the flash is produced as easily as in the air. My attention was drawn to this fact by a letter from the well-known naturalist the Rev. Reginald Gatty. I at once tried the experiment and found the fact to be as my correspondent stated. Not only so, but the smell was produced as well as the flash.

With the desire to get further light on the subject,

I consulted the great experimental physicist, my friend Sir James Dewar, in his laboratory at the Royal Institution. He told me that the late Professor Tyndal used to exhibit the production of flame by the friction of two pieces of quartz in his lectures on heat, but made use of a very large and rough crystal of quartz (rock-crystal) and rubbed its rough surface with another large crystal. Tyndal's note on the subject in his lecture programme was as follows (Juvenile Lectures on Heat, 1877–78): "When very hard substances are rubbed together light is produced as well as heat." Sir James Dewar kindly showed me the crystals used by Tyndal, the larger was 16 inches long and 4 or 5 inches broad. We repeated the experiment in the darkened lecture room, and obtained splendid flashes. The same smell is produced when rock-crystal is used as when flint or quartz pebbles are rubbed together. All three are the same chemical body, namely, silica (oxide of silicon). We also found that when the crystals were bathed with water or (this is a new fact) with absolute alcohol, the same flashing was produced by the friction of one against the other.

Later, with the kind assistance of Mr. Herbert Smith, of the mineral department of the Natural History Museum, I examined, with a spectroscope, the flash given by two quartzite pebbles when rubbed together. No distinctive lines or bands were seen; only a "continuous" spectrum, showing that the temperature produced was not high enough to volatilize the silicon. I also examined some pebbles of another very hard substance—nearly as hard as silica (rock-crystal, quartz, and flint). This was what is called "corundum," the massive form of "emery powder" (oxide of aluminium). By grinding two of these corundum pebbles with very great pressure one against the other (using much greater pressure than is needful in the case of quartz), I obtained flashes of light. It was not known previously that any pebbles except those of silica would give flashes of light when rubbed together. A smell resembling that given out by rubbed quartz, but fainter, was observed.

Those are the facts—new to me and to many others—about this curious subject. The flashing under water is a very remarkable thing. I cannot say that I am yet satisfied as to the nature of the flash. A simple explanation of the result obtained, when two dry pebbles are rubbed together in the air, is that crushed particles of the quartz or of the corundum are heated by the heavy friction to the glowing point. But this does not accord with the fact that submergence in a liquid does not interfere with the flashing. The rise of temperature would certainly be checked by the liquid. And the curious smell produced is in no way explained.

The breaking of crystals is in many instances known to produce a flash of light. Thus a lump of loaf sugar broken in the dark gives a faint flash of blue light, as anyone can see for himself immediately on reading this. White arsenic crystals also, when broken by shaking the liquid in which they have formed, give out flashes of light. Some rare specimens of diamond, when rubbed in the dark with a chamois leather, glow brightly. The well-known mineral called Derbyshire spar, "Blue John," or fluoride of calcium, when heated to a point much below that of a red-hot iron, "crackles" and glows briefly with a greenish light. The crystals of phosphate of lime, called apatite, and a number of other crystals have this property. But there is no record of any peculiar smell accompanying the flashes of light. It is still a matter open to investigation as to whether the flashing of pieces of quartz and rock-crystal when rubbed together with heavy pressure is of the nature of the flashing of the heated crystals of other minerals, or whether there is any chemical action set up by the friction—an action which is certainly suggested by the very peculiar smell produced. Since the flashing can be produced under water and other liquids, it should be easy to obtain some evidence as to the chemical nature of the flame—whether acid or alkaline, whether capable of acting on this or that reagent dissolved in the water, and whether setting free any gas of one kind or another.

Any one of my readers who chooses can produce the wonderful orange-coloured flame by rubbing two quartz or flint pebbles together in the dark, and can have the further gratification of producing with the utmost ease the mysterious and weird phenomenon of a flame under water, and may, perhaps, by further experiment, explain satisfactorily this unsolved marvel which has haunted some of us since childhood.

CHAPTER IX

AMBER

AMBER is not unfrequently picked up among the pebbles of the East Coast. I once picked up a piece on the beach at Felixstowe as big as a turkey's egg, thinking it was an ordinary flint-pebble and intending to throw it into the sea, when my attention was arrested by its extraordinary lightness, and I found that I had got hold of an unusually large lump of amber. There is a locality where amber occurs in considerable quantity. It is a long way off—namely, the promontory called Samland near Königsberg on the Prussian shore of the Baltic. There it occurs with fossil wood and leaves in strata of early Tertiary age, deposited a little later than our "London clay." It used to be merely picked up on the shore there until recent times, when "mining" for it was started. From this region (the Baltic coast of Prussia) amber was carried by the earliest traders in prehistoric times to various parts of Europe. Their journeyings can be traced by the discovery of amber beads in connexion with interments and dwelling-places along what are called "amber routes" radiating from the amber coast of Prussia. To reach the East Coast of England the bits of amber would have to be carried by submarine currents. Amber travels faster and farther than ordinary stones, on account of its lightness. What has been held to be amber is found, also embedded in ancient Tertiary strata, in small quantity in France, in Sicily, in Burma, and in green sand (below the chalk) in the United States. The Sicilian amber (called "Simetite") was not known to the ancients: it is remarkable for being "fluorescent," as is also some recently discovered in Southern Mexico. But it is possible that chemically these substances are not quite the same as true amber. Amber is a fossil resin or gum, similar to that exuded by many living trees, such as gum-copal. It has been used as an ornament from prehistoric times onwards, and was greatly valued by the Egyptians, Greeks, and Romans, and by our Anglo-Saxon ancestors, not only for decorative purposes, but as a "charm," it being supposed to possess certain magical properties.

Amber (it is generally believed) comes slowly drifting along the sea bottom to the Suffolk shore from the Baltic. Lumps as big as one's fist are sometimes picked up here. The largest pieces on record found on the Baltic shore, or dug out of the mines there, are from 12 to 18 lb. in weight, and valued at £1000. A party sent by the Emperor Nero brought back 13,000 lb. of amber from the Baltic shores to Rome. The bottom currents of seas and oceans, such as those which possibly bring amber to our shores, are strangely disposed. The Seigneur of Sark some fifty years ago was shipwrecked in his yacht near the island of Guernsey; he lost, among other things, a well-fastened, strongly-made chest, containing silver plate. It was found a year later in deep water off the coast of Norway and restored to him! In the really deep sea, over 1000 fathoms down, there are well-marked broad currents which may be described as rivers of very cold water (only four degrees or so above freezing-point). They flow along the deep sea bottom and are sharply marked off from the warmer waters above and to the side. Their inhabitants are different from those of the warmer water. They are due to the melting of the polar ice, the cold water so formed sinking at once owing to its greater density below the warmer water of the surface currents. These deep currents originate in both the Arctic and Antarctic regions, and the determination of their force and direction, as well as of those of other ocean currents, both deep and superficial, such as the warm "Gulf Stream," which starts from the Gulf of Mexico, and the great equatorial currents, is a matter of constant study and observation, in which surveying ships and skilled observers have been employed.

Amber has not only been valued for its beauty of colour—yellow, flame-colour, and even deep red and sometimes blue—for its transparency, its lightness, and the ease with which it can be carved, but also on account of certain magical properties attributed to it. Pliny, the great Roman naturalist of the first century A.D., states that a necklace of amber beads protects the wearer against secret poisoning, sorcery, and the evil eye. It is first mentioned by Homer, and beads of it were worn by prehistoric man. Six hundred years B.C., a Greek observer (Thales) relates that amber when rubbed has the power of attracting light bodies. That observation is the starting-point of our knowledge of electricity, a name derived from the Greek word for amber, "electron." In Latin, amber is called "succinum." By heating in oil or a sand-bath, amber can be melted, and the softened pieces squeezed together to form larger masses. It can also be artificially stained, and cloudy specimens are rendered transparent by heating in an oil-bath.

Amber is the resinous exudation of trees like the

"Copal gum" of East Africa and the "Kauri resin" or "Dammar" of New Zealand. Both of these products are very much like amber in appearance, and can be readily mistaken for it. The trees which produced the amber of the Baltic were conifers or pine trees, and flourished in early Tertiary times (many millions of years ago). Their leaves, as well as insects of many kinds, which have been studied and named by entomologists, are found preserved in it. There is a very fine collection of these insects in the Natural History Museum in London. It is probable that more than one kind of tree produced the amber-gum, and that its long "fossilization" has resulted in some changes in its density and its chemical composition. The East African copal is formed by a tree which belongs to the same family as our beans, peas, and laburnum. It is obtained when freshly exuded, but the best kind is dug by the negroes out of the ground, where copal trees formerly grew and have left their remains, so that copal, like amber, is to a large extent fossilized. The same is true of the New Zealand dammar or kauri gum, which is the product of a conifer called "Agathis australis," and is very hard and amber-like in appearance. Chemically amber, copal, and dammar are similar to one another but not identical. Amber, like the other two, has been used for making "varnish," and the early Flemish painters in oils, as well as the makers of Cremona violins, made use of amber varnish.

A medicament called "eau de luce" was formerly used, made by dissolving one of the products of the dry distillation of amber (called "oil of amber") in alcohol. Now, however, amber is used only for two purposes—besides decoration—namely, for the mouthpieces of pipes and cigar tubes and for burning (for amber, like other resins, burns with a black smoke and agreeable odour) as a kind of incense (especially at the tomb of Mahomet at Mecca). These uses are chiefly Oriental, and most European amber now goes to the East. In China they use a fine sort of amber, obtained from the north of Burma. The use of amber as a mouthpiece is connected with its supposed virtues in protecting the mouth against poison and infection. It is softer than the teeth, and therefore pleasant to grip with their aid; but as a cigar or cigarette tube it is disadvantageous, as it does not absorb the oil which is formed by the cooling of the tobacco smoke passing along it, but allows it to condense as an offensive juice.

Forty years ago an old lady used to sit in the doorway of her timber-built cottage in the village of Trimley (where there are the churches of two parishes in one churchyard), smoking a short clay pipe and carving bits of amber found on the Suffolk beach into the shape of hearts, crosses, and beads. She would carve and polish the amber you had found yourself whilst you joined her in a friendly pipe. You were sure in those days of the genuine character of the amber, jet, and agate sold as "found on the beach." Nowadays these things, as well as polished agates and "pebbles from the beach," are, I am sorry to say, manufactured in Germany, and sent to many British seaside resorts, like the false coral and celluloid tortoise-shell which, side by side with the genuine articles, are offered by picturesque Levantines to the visitors at hotels on the Riviera, and even in Naples itself. Nevertheless, genuine and really fine specimens of amber picked up on the beach and polished so as to show to full advantage their beautiful colour and "clouding" can still be purchased in the jeweller's shop at Aldeburgh on the

Suffolk coast near the great pebble beach of Orfordness.

There are difficulties about using the word "amber" with scientific precision. The fossil resins which pass under this name in commerce, and are obtained in various localities, including the Prussian mines on the Baltic, are undoubtedly the product of several different kinds of trees, and, from the strictly scientific chemical point of view, they are mixtures in varying proportions of different chemical substances. The merchant is content with a certain hardness (which he tests with a penknife), transparency, and colour, and also attaches great importance to the test of burning a few fragments in a spoon, when, if the material is to pass as "amber," it should give an agreeable perfume. Scientifically speaking, "amber" differs from other "resins," including copal, in having a higher melting point, greater hardness, slighter solubility in alcohol and in ether, and in containing "succinic acid" as an important constituent, which the other resins, even those most like it, do not. True amber thus defined is called "succinite," but several other resins accompany it even as found in its classical locality—the Baltic shore of Prussia—and, owing to their viscid condition before fossilization, may have become mixed with it. One of these is called "gedanite," and is used for ornamental purposes. It is more brittle than amber, and contains no succinic acid. It is usually clear and transparent, and of a pale wine-yellow colour.

It is not possible to be certain about the exact nature of what appears to be a "piece of amber" thrown up on the seashore, without chemical examination. A year or two ago a friend brought to me a dark brownish-yellow-coloured piece of what looked like amber, which (so my friend stated) had been picked up on the shore at Aldeburgh. It was as big as three fingers of one's hand, very transparent and fibrous-looking, owing to the presence of fine bubbles in its substance arranged in lines. I found an exactly similar piece from the same locality in the collection of the Natural History Museum. It was labelled "copal," and, I suppose, had been chemically ascertained to be that resin and not "amber," or, to use the correct name, "succinite." How either of these pieces got into the North Sea it is difficult to say. Though the "copal" of commerce is obtained from the West Coast of Africa, it may occur (though I have not heard that it does) associated with true amber in Prussia. A fossilized resin very similar to copal is found in the London clay at Highgate and elsewhere near London, and is called "copalite." It is possible, though not probable, that the bits of amber found on our East Coast beaches are derived from Tertiary beds, now broken up and submerged in the North Sea, and do not travel to us all the way from the Baltic.

CHAPTER X

SEA-WORMS AND SEA-ANEMONES

LET us now leave the beach-pebbles and go down on to the rocks at low tide in order to see some of the living curiosities of the seashore. There are some seaside resorts where, when the tide goes down, nothing is exposed but a vast acreage of smooth sand, and here the naturalist must content himself with such spoils as may be procured by the aid of a shrimping-net and a spade. Wading in the shallow water and using his net, he will catch, not only the true "brown shrimp," but other shrimp-like creatures, known as "crustacea"—a group which includes also the lobsters, hermit-crabs, true crabs, and sand-hoppers, as well as an immense variety of almost microscopic water-fleas.

He will also probably catch some of the stiff, queer little "pipe-fish," which are closely related to the little creatures known as "sea-horses." Pipe-fish are very sluggish in movement, almost immobile, whilst the "sea-horse" or hippocampus—only to be taken by the dredge amongst corallines in deep water on rocky bottoms (as, for instance, in the Channel Islands)—goes so far as to curl his tail, like a South American monkey, round a stem of weed and sit thus upright amidst the vegetation. Even when disturbed he merely swims very slowly and with much dignity in the same upright position, gently propelled by the undulating vibratory movement of his small dorsal fin. The male in both pipe-fish and sea-horses is provided with a sac-like structure on the ventral surface in which he carries the eggs laid by the female until they are hatched.

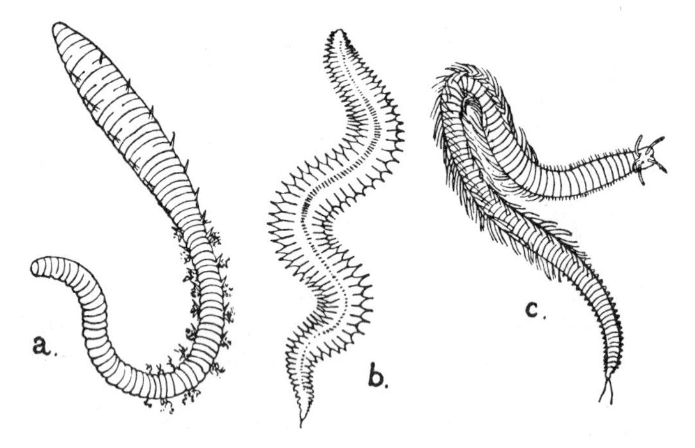

FIG. 4.—British Marine Worms or Chætopods.

a, Arenicola piscatorum. Lug-worm largely used for bait by sea-fishermen. It burrows in sea-sand and clay as the earth-worm does in soil. Half the natural size, linear.

b, Nephthys margaritacea, actively swimming. It also burrows in the sea-sand. Natural size.

c, Eunice sanguinea, a very handsome marine worm (often used for bait) which lives in clefts in the submarine rocks and also swims actively. The numerous filaments on the sides of the ringed body are the gills of a rich blood-colour. The figure is one-third of the natural size, linear.

The shrimper will probably catch also some very young fish fry—including young flat-fish about 2 inches long. If he explores the exposed surface of sand near the low-tide limit, he will find a variety of indications of burrowing animals hidden beneath. Little coiled masses like the "castings" of earth-worms are very abundant in places, and are produced by the fisherman's sand-worm, or "lug-worm" (Fig. 4, a). A vigorous digging to the depth of a foot or two will reveal the worm itself, which is worth bringing home in a jar of sea-water in order to see the beautiful tufts of branched gills on the sides of the body, which expand and contract with the flow of bright red blood showing through their delicate walls. Other sand-worms, from 2 to 6 inches long, will at the same time be turned up,—worms which have some hundred or more pairs of vibrating legs, or paddles, arranged down the sides of the body, and swim with a most graceful, serpentine curving of the mobile body (Fig. 4, b). These sea-worms are but little known to most people, although they are amongst the most beautifully coloured and graceful of marine animals. Hundreds of different kinds have been distinguished and described and pictured in their natural colours. Each leg is provided with a bundle of bristles of remarkable shapes, resembling, when seen under a microscope, the serrated spears of South Sea Islanders and mediaeval warriors. These worms usually have (like the common earth-worm) red blood and delicate networks of blood-vessels and gills (Fig. 4, c), whilst the head is often provided with eyes and feelers. They possess a brain and a nerve-cord like our spinal cord, and from the mouth many of them can suddenly protrude an unexpected muscular proboscis armed with sharp, horny jaws, the bite of which is not to be despised. These "bristle-worms," or "chætopods," as they are termed by zoologists, are well worth bringing home and observing in a shallow basin holding some clean sea-water.

At many spots on our coast (*e.g.* Sandown, in the Isle of Wight, and the Channel Islands) rapid digging in the sand at the lowest tides will result in the capture of sand-eels, a bigger and a smaller kind, from 1 foot to

6 inches in length. These are eel-shaped, silvery fish, which swim near the shore, but burrow into the soft sand as the tide recedes. They are excellent eating. We used at Sandown to make up a party of young people to dig the smaller "sand-eels," or "sand-launce." The agility and rapid disappearance of the burrowing fish into the sand when one thought one had safely dug them out, rendered the pursuit difficult and exciting. Then a wood fire on the beach, a frying-pan, fat, flour, and salt were brought into operation, and the sand-eels were cooked to perfection and eaten.

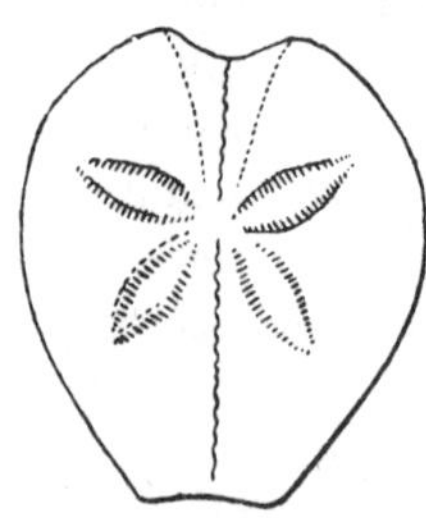

FIG. 5.—The shell of the Heart-urchin (Spatangus purpureus) with its spines rubbed off. One-fourth the actual diameter.

Some of the marks or small heaps of sand on the flats exposed at low tide are characteristic of certain shell-fish. The "razor-fish" (Fig. 19,b)—a very much elongated clam, or mussel, with astonishing powers of rapid burrowing—leaves a hole on the surface like a keyhole, about an inch long. It can be dug up by an energetic spadesman, but a spoonful of common salt poured over the opening of its burrow will cause it to suddenly shoot out on to the surface, when it may be picked up, and the hunter spared any violent exertion. The curious heart-urchin (Fig. 5), as fragile as an egg-shell, and covered with long, closely-set spines like a brush, is often to be found burrowing in the sand, as well as the transparent, pink-coloured worm known as Synapta, in the skin of which are set thousands of minute calcareous anchors hinged to little sculptured plates. These burrowers swallow the sand and extract nutriment from stray organic particles mixed in it.

The mere sand-flat of the low tide is not a bad hunting ground; but the rock pools, often exposed when the tide is out, and the fissures in the rocks and the under surfaces of slabs of rock revealed by turning them over—are the greatest sources of varied delight to the sea-shore naturalist. It is well to take a man with you on to these rocks to carry your collecting bottles and cans, and to turn over for you the larger slabs of loose stone, weighing as much as a couple of hundredweight. The most striking and beautiful objects in these rock pools are the sea-anemones (Fig. 6 and Frontispiece). They present themselves as disk-like flowers from 1 to 5 inches in diameter, with narrow-pointed petals of every variety of colour, set in a circle around a coloured centre. The petals are really hollow tentacles distended with sea-water, and when anything falls on to them or touches them they contract and draw together towards the centre. The centre has a transverse opening in it which is the mouth, and leads into a large, soft-walled stomach, separated by its own wall from a second spacious cavity lying between that wall and the body wall, and sending a prolongation into each tentacle. The stomach opens freely at its deep end into this second "surrounding" chamber, which is divided by radiating cross walls into smaller partitions, one corresponding to each tentacle. The nourishing results of digestion, and not the food itself, pass from the stomach into the subdivided or "septate" second chamber. There is thus only one cavity in the animal, separable into a central and a surrounding portion.

In this respect—in having only one body cavity—sea-anemones and the coral-polyps and the jelly-fishes and the tiny freshwater polyp or hydra, and the marine compound branching polyps like it—agree with one another and differ from the vast majority of animals, such as worms, sea-urchins, star-fishes, whelks, mussels, crustaceans, insects, spiders and vertebrates (which last include fish, reptiles, birds, and mammals). These all have a second chamber, or body cavity, quite shut off from the digestive cavity and from the direct access of water and food particles. This second distinct chamber is filled with an animal fluid, the lymph, and is called the "Cœlom" (a Greek word meaning a cavity). These higher animals, which possess a cœlom as well as a gut, or digestive cavity, are called "Cœlomata," or "Cœlomocœla," in consequence; whilst the sea-anemones, polyps, and jelly-fish form a lower grade of animals devoid of cœlom, but having the one cavity, or gut, continued into all parts of the body. Hence they are called "Cœlentera," or "Enterocœla," words which mean that the cavity of their bodies (Greek *cœl*) is made by an extension of the gut, or digestive cavity (Greek *enteron*). The higher grade of animals—the Cœlomocœla—very usually have a vascular system, or blood-vessels and blood, as well as a cœlom and lymph, and quite independent of it; also some kind of kidneys, or renal excretory tubes. Neither of these are possessed by the sea-anemones and their allies—the Enterocœla—but they have, like higher animals, a nervous system and also large ovaries and spermaries on the walls of their single body cavity, which produce their reproductive germs. These pass to the exterior, usually through the mouth, but sometimes by rupture of the body wall.

All "one-cavity" animals, the Enterocœla or Cœlentera, produce peculiar coiled-up threads in their skin in great quantity—many thousands—often upon special warts or knobs. These coiled-up threads lie each in a microscopic sac; they are very delicate and minute and carry a virulent poison, so that they are "stinging" threads. Excitement of the animal, or mere contact, causes the microscopic sac to burst, and the thread to be violently ejected. The sea-anemones, jelly-fish, and polyps feed on fresh living animals, small fish, shrimps, etc., and catch their prey by the use of these poisonous threads. Some jelly-fish have them big enough to act upon the human skin, and bathers are often badly stung by them. The commonest jelly-fish do not sting, but where they occur a few of the stinging sort are likely to occur also. Even some sea-anemones can sting one's hand with these stinging threads. One sea-anemone (known as "Cerianthus"), occasionally taken in British waters, makes for itself a leathery tube by the felting of its stinging threads, and lines its long burrow in the sand below tidal exposure in this way.

The sea-anemones are very hardy, and they are wonderfully varied and abundant on our coasts. Some sixty years ago a great naturalist, who loved the sea-shore and its rock-pools enthusiastically, Mr. Philip Henry Gosse, father of Mr. Edmund Gosse, the distinguished man of letters, described our British sea-anemones, and gave beautiful coloured pictures of them. One of these I have taken for the frontispiece of this volume, and some of the outline figures of marine animals in these chapters are borrowed from a marvelously complete and valuable little book by him—now long out of print—entitled "Marine Zoology." His books—of high scientific value—and his example, made sea-anemones "fashionable." London ladies kept marine aquariums in their drawing-rooms stocked with these beautiful flowers of the sea. They were exhibited in quantity at the Zoological Gardens in Regent's Park, and it is by no means a creditable thing to our London zoologists that neither these nor other marine creatures are now to be seen there. At a later date public marine aquaria were started with success in many seaside towns,—Brighton, Scarborough, Southport, etc.—and a very fine one was organized in Westminster and another at the Crystal Palace. It is an interesting and important fact, bearing on the psychology of the British people, that most of these charming exhibitions of strange and beautiful creatures from the depths of the sea were very soon neglected and mismanaged by their proprietors; the tanks were emptied or filled with river water, and the halls in which they were placed were re-arranged for the exhibitions of athletes, acrobats, comic singers, and pretty dancers. These exhibitions are often full of human interest and beauty—but I regret the complete disappearance of the fishes and strange submarine animals. I have some hope that before long we may, at any rate in the gardens in the Regent's Park, see really fine marine and fresh-water aquaria established, more beautiful and varied in their contents than those of earlier days.

There are four kinds of sea-anemones which are abundant on our coast. They adhere by a disk-like base to the rocks and large stones, and have the power of swelling themselves out with sea-water (as have many soft-bodied creatures of this kind), with all their tentacles expanded. They have, in that condition, the shape of small "Martello" towers, with their adhesive disk below and the mouth-bearing platform above, fringed by tapering fingers; and they can, on the other hand, shrink to a fifth part of their expanded volume, drawing in and concealing their tentacles, which are in some kinds perforated at the tip. One common on the rocks at Shanklin and other parts of our South Coast, but

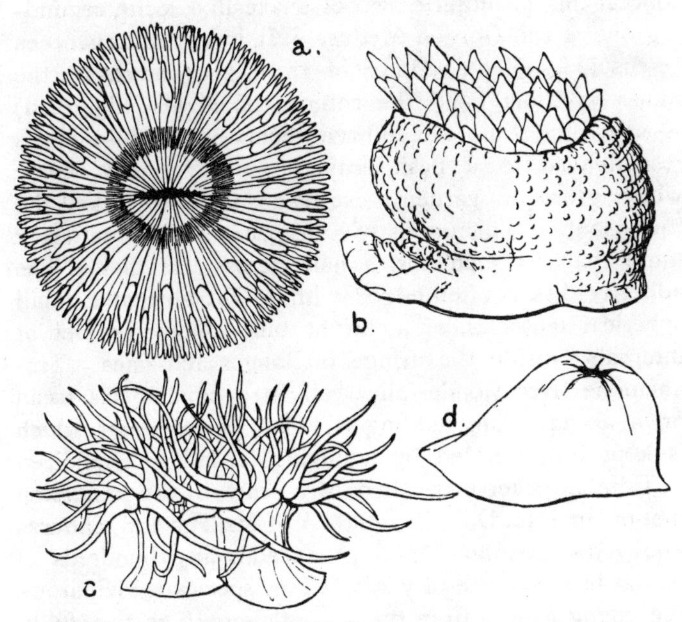

FIG. 6.—British Sea-Anemones.

a, Sagartia bellis, the daisy anemone, viewed from above when fully expanded.

b, Bunodes crassicornis, half expanded ; side view.

c, Anthea cereus. The tentacles are pale apple-green in colour, tipped with mauve, and cannot be completely retracted.

d, Actinia mesembryanthemum. The disk of tentacles is completely retracted. This is the commonest sea-anemone on our South Coast, and is usually maroon colour, but often is spotted like a strawberry.

not on the East Coast, has very abundant, long, pale green tentacles, which are tipped with a brilliant peach colour, and it is peculiar in not being able to retract or conceal this beautiful crown of snake-like locks, reminding one of the Gorgon Medusa. It is known as Anthea cereus (Fig. 6, c). Many of them are known by the name "Actinia," and the commonest of all (Fig. 6, d) is called "Actinia mesembryanthemum," because of its resemblance to a fleshy-leaved flower of that name which grows on garden rockeries—sometimes called the "ice-plant." This one is of a deep maroon colour, rarely more than an inch and a half across the disk. The adhesive disk is often edged with bright blue, and small spherical tentacles, of a bright blue colour, are set at intervals outside the fringe of longer red ones. This anemone lives wonderfully well in a small glass basin or in an aquarium holding a gallon of sea-water, which is kept duly aerated by squirting it daily. One lived in Edinburgh for more than fifty years, in the possession first of Sir John Dalyell, and then of Mr. Peach. She was known as "Granny," and produced many hundreds of young in the course of years. This species is viviparous, the young issuing from the parent's mouth as tiny fully-formed sea-anemones, which immediately fix themselves by their disks to the glass wall of their habitation. Anemones kept thus in small aquaria have to be carefully fed; bits of the sea mussel (of course, uncooked) are the best food for them. This and many other kinds are not absolutely stationary, but can very slowly crawl by means of muscular movements of the adhesive disk. There are kinds of sea-anemones known which spend their lives floating in the ocean; they are thin and flat. Others adhere to the shells of hermit crabs and even to the big claws of some crabs, and profit by the "crumbs" of food let fall by the nippers of their host.

A very handsome and large sea-anemone is common on the East Coast, and is known as "crassicornis" (its generic name is Bunodes). When distended it measures as much as 4 inches across (Fig. 6, b). I have one at this moment before me, expanded in a bowl of sea-water. The tentacles are pale green or grey, banded with deep red, and the body is blotched with irregular patches of red, green, and orange. It attaches fine pebbles and bits of shell to the surface of the body.

CHAPTER XI

CORAL-MAKERS AND JELLY-FISH

A VERY beautiful kind of sea-anemone (common at Felixstowe) is the Daisy or Sagartia troglodytes, (Fig. 6, a), which has a very long body attached to a rock or stone far below the sandy floor of the pool, on the level of which it expands its thin, long, ray-like tentacles, coloured dark brown and white, and sometimes orange-yellow. As soon as you touch it it disappears into the sand, and is very difficult to dig out. The most beautifully coloured of all sea-anemones are the little Corynactids (half an inch across), which you may find dotted about like jewels, each composed of emerald, ruby, topaz, and creamy pink and lilac, on the under surface of slabs of rock at very low tide in the Channel Islands. One of the most puzzling facts in natural history is that these lovely little things live in the dark. No eye, even of fish or crab, has ever seen what you see when you turn over that stone. It is a simple demonstration of the truth of the poet Gray's statement, that many a gem of purest ray serene is concealed in the dark, unfathomed depths of ocean! A splendid anemone is the Weymouth Dianthus (see the frontispiece of this volume), so named because it is dredged up in Weymouth Bay. It is often six inches long, and has its very numerous, small tentacles arranged in lobes, or tufts, around the mouth. It is either of a uniform bright salmon-yellow colour or pure white. When kept in an aquarium it fixes itself by its disk on the glass wall, and often, as it slowly moves, allows pieces of the disk to become torn off and remain sticking to the glass. These detached pieces develop tentacles and a mouth, and grow to be small and ultimately full-sized Weymouth anemones.

If the disk were spread out and gave rise to little anemones without tearing—so that they remained in continuity with the parent—we should get a composite or compound animal, made up of many anemones, all connected at the base. This actually happens in a whole group of polyps resembling the sea-anemones. They grow into "stocks," "tree-like" or "encrusting" masses, consisting of hundreds and even thousands of individuals, each with its mouth and tentacles, but with their inner cavities and bases united. These are the "coral polyps," or "coral-insects" of old writers, of so many varied kinds. One further feature of great importance in a "coral" is the production of a hard deposit of calcite, or limestone, which is thrown down by the surface of the adhesive disk, and is also formed in deep, radiating "pockets," pushed in to the soft animal from the disk. The hard deposit of calcite is continuous throughout the "stock," or "tree," and when the soft sea-anemone-like animals die, the hard, white matter is left, and is called "coral." Very commonly this white coral shows star-like cups on its surface, which correspond to the lower ends or disks of the soft sea-anemone-like creatures which deposited the hard coral. In a less common group (represented commonly on our coast by the so-called "Dead men's fingers" found growing on the overhanging edges of low-tide rocks) the hard coral material does not form cups for the minute sea-anemones which secrete it, but takes the form of a supporting central or axial rod (sea-pens), or branched tree (sea-bushes), upon which the fleshy mass of polyps are tightly set. This is the case with the precious red and pink coral of the Mediterranean (which is now being "undersold" actually in the Mediterranean markets by a similar red coral from Japan, usually offered as the genuine article, which it is not!).

On the British coast you do not, as a rule, find coral-forming polyps. A small kind, consisting of two or three yellow and orange-red anemone-polyps united and producing a small group of hard calcite cups (Caryophyllia and Balanophyllia) is not uncommon at Plymouth at a few fathoms depth. But you have to go to the Norwegian fiords or else far out to sea where you have 300 fathoms of sea-water in order to get really luxuriant white corals—the beautiful Lophohelia (Fig. 3, p. 9), which I used to dredge in the Nord Fiord near Stavanger, as branching, shrub-like masses of a foot cube in area, each white marble cup standing out from the stem, an inch long and two-thirds of an inch across, and the stems giving support to a whole host of clinging growths (among them Rhabdopleura!) and sheltering wonderful deep-water worms and starfish.

But these, beautiful as they are, are nothing, so far as mass and dominating vigour of growth are concerned, in comparison with the reef-building corals of the warm seas of the tropics. There these lime-secreting conglomerated sea-anemones separate annually hundreds of tons of solid calcite per square mile of sea bottom from the sea-water, and build up reefs, islands, and huge cliffs of coral rock. They get the calcite—as do calcareous seaweeds and shell-making clams, oysters, whelks, and microscopic chalk-makers—from the sea—the water of the sea which always has it ready in solution for their use. And the sea gets it from the rivers and streams which wear away and dissolve the old limestone deposits now raised into mountain chains, as well as by itself dissolving again in due course what living creatures have so carefully separated from it. Sea water or fresh water with a little carbonic acid gas dissolved in it dissolves limestone and chalk—it becomes what we call "hard." Neutralize the dissolved carbonic acid (as is done in the well-known Clark's process for softening water), and down falls the dissolved calcite as a fine white sediment. These alternating processes of solution and "precipitation" are always going on in the waters of the earth and sea.

The name "jelly-fish" has reference to the colourless, transparent, soft, and jelly-like substance of the bodies of the animals to which it is applied. There are a number of marine animals, besides the common jelly-fish, belonging to different classes, which are glass-like in transparency and colourless—so as to be nearly or quite invisible in clear water, and some, too, occur in fresh waters (larvæ of gnats, notably of the plume-horned gnat Corethra). The transparency of these animals serves them in two different ways—some are enabled by it to escape from predatory enemies; others, on the contrary, are enabled to approach their own prey without being observed. The latter was obviously the case with the little fresh-water jelly-fish which appeared in great abundance some years ago in the lily tank in Regent's Park. The water was full of small water-fleas (minute crustacea), and the little jelly-fish, if removed from the tank and placed in a tall glass jar filled with the tank water, spent its whole time in swimming upwards to the surface by the alternate contraction and expansion of its disk-like body, and then dropping gently through the full length of the jar to the bottom, when it would again mount. On the downward journey—owing to its transparency—it would encounter unsuspecting, jerkily-moving water-fleas, unwarned by any shadow cast by the impending glass-like monster of half an inch in breadth slowly approaching from above; and as soon as they touched it they were paralysed (by microscopic poison-threads like those of the sea-anemones), and were grasped and swallowed by the mobile transparent proboscis (like that of an elephant, though certainly smaller, and having the mouth opening at its end, instead of a nostril), which hangs from the centre of the disk-like jelly-fish.[1]

There are some glass-like transparent creatures, including some small fishes, which live at 500 fathoms depth and a good deal deeper on the sea bottom. We know that the sun's light does not penetrate below 200 fathoms, so that one is led to ask—What is the good of being transparent if you live at the bottom of the sea, at a greater depth than this? There is also a very beautiful prawn, which I dredged in Norway in 200 fathoms, which looks like a solid piece of clearest, colourless glass. And then there are some very beautiful little stalked creatures (called Clavellina), fixed to the under-side of rocks in the tidal zone, which are absolutely like drops of solid glass an inch long. One cannot easily imagine how colourless transparency can be of "life-saving value" to these varied inhabitants of the dark places of the sea bottom—any more than we

[1] See "Science from an Easy Chair" (First Series, 1910), p. 60, for a further account and figure of the freshwater jelly-fish.

can assign any life-saving value to the brilliant, gem-like colouring of some of the sea-anemones which live in the dark on the under-surface of rocks.

The most probable view of the matter is that neither the colourless transparency of the one set nor the brilliant colouring of the other has any value; it just happens to be so, and is not harmful. So, for instance, some crystals are colourless, some blue or green or yellow or red, without any advantage to them! On the other hand, we know that a large number of the animals which live in the dark unfathomed depths themselves produce light, that is to say, are phosphorescent, and it seems probable that at great depths, though there is no sunlight, the sea bottom is illuminated—we can only vaguely guess to what degree—by the strange living lanterns—fish, crustaceans, worms, and even microscopic creatures—which move about in quest of their food, carrying their own searchlight with them. Another suggestion is that the eyes of these inhabitants of the dark may be more sensitive than our own, and even be affected by rays invisible to us. This, however, is not probable, since whilst there are among them some with enormous eyes, we find that at the greatest depths (2 to 4 miles) even the fishes have no eyes at all, and at a depth of a mile there are many shrimp-like creatures in which the eyes have been completely transformed into peculiar "feelers," or otherwise aborted. So that we cannot suppose there is a possibility of developing the eye of the dwellers in deep-sea darkness to a degree of sensitiveness greatly beyond that of terrestrial animals. A limit of obscurity is reached at which it is of no use having an eye at all, and eyes cease to have life-saving value, and accordingly are not maintained by natural selection.

The transparency and colourlessness of marine animals which float near the surface is, on the other hand, obviously useful, and to this group our jelly-fishes belong. Not only do they escape observation by their transparency and general absence of colour, but some actually have a blue transparent colouring which blends with the blue colour of the sea. Such are the gas-holding, bladder-like sac as large as your fist called the "Portuguese man-of-war," and the little sailing Velella, both of which float, and even protrude above the surface, so as to catch the wind. Others are only semi-transparent, and others are marked with strong red, brown, or yellow streaks. Many of the smallest kinds of jelly-fish have eyes which are bright red in colour.

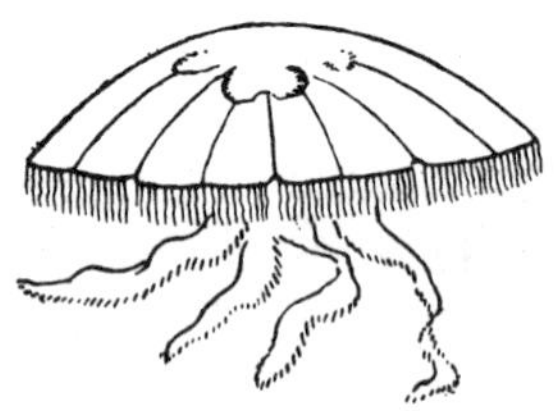

FIG. 7.—A common British Jelly-fish.

Aurelia aurita, usually as large as a breakfast-plate and often larger.

The animals to which the name "jelly-fishes" is now more or less strictly applied are (as that fine zoologist Aristotle knew) in their structure closely similar to the sea-anemones, but even simpler. They are called the Medusæ by naturalists. Their disk-like bodies are largely formed by a jelly-like material, on the surface of which are stretched delicate transparent skin, nerves, and delicate muscles, whilst in the middle of the disk, on the surface which faces downwards as the creature floats, is the mouth, leading into a relatively small pouched cavity excavated in the jelly, from which a delicate system of canals is given off, and radiates in the jelly of the disk. There is, as in the sea-anemones, only one continuous cavity. The edge of the disk is beset with fine, sensitive tentacles, sometimes many feet in length, and the lips of the mouth are often drawn out into a sort of depending trunk, or into four large tapering lobes or lips of jelly, which, with the longer tentacles, are used for seizing prey. The commonest jelly-fish on our coast—so common as to be "the" jelly-fish *par excellence*—is often to be seen left on the sands by the receding tide or slowly swimming in quiet, clear water at the mouth of a river in enormous numbers. It is known as "Aurelia" (Fig. 7). It is as big as a cheese-plate, and the four pouches connected with the stomach are coloured pink or purple, and appear in the middle of the circular plate of jelly, like a small Maltese cross. The reproductive particles (germ-cells and sperm-cells) are produced in that coloured region, and escape by the mouth. There is a fringe of fine, very short tentacles round the edge of the disk, and they, as well as the great lobes of the mouth, are provided with innumerable coiled-up stinging hairs or "thread-cells," similar to those of the sea-anemones, which led Aristotle to call both groups "sea-nettles." Eight stalked eyes are set at equal intervals around the disk.

Usually accompanying the floating crowd of the common and abundant Aurelia are a few specimens of a very unpleasant kind of Medusa of a turbid appearance, often called "slime balls" by fishermen, from six inches to a foot in diameter. It is known to naturalists by the name "Cyanæa capillata." The tentacles on the edge of the disk of this kind of jelly-fish are very long and elastic, stretching to several feet, even yards, in length, and are provided with very powerful stinging hairs. The tentacles not infrequently become coiled around the body of a bather; the stinging hairs are shot out of the little sacs in which they are rolled up, and the result may be very painful to the person stung in this way and even dangerous. There are two other common large jelly-fish on the English coast, one called "Chrysaora" (Fig. 8), with a wheel-like pattern of brown pigment on the disk, and the other with the mouth lobes very large and bound together like a column.

FIG. 8.—A common British Jelly-fish.
Chrysaora hysoscella, usually twice as big as the figure.

The common Aurelia is remarkable for the fact that the young which hatch from its eggs attach themselves to stones and rocks on the sea bottom, and grow into little white tube-like polyps, about half an inch long, quite unlike their parent, with a crown of small tentacles surrounding the mouth, whilst they are fixed by the opposite end of the body. Then a very curious thing happens. The little polyp becomes nipped at intervals across its length, so that it looks like a pile of saucers — a dozen or more. And then the top saucer swims away as a minute jelly-fish, the next follows, and so on, so that, in the course of an hour or two, the whole pile separates into a number of freely swimming young, each of which gradually grows into a full-sized Aurelia. I have only once had the chance of witnessing this beautiful sight, and that was many years ago in a tank at the Zoological Gardens (they have no such tanks now), where the polyp-like young (called "Hydra tuba") spontaneously put in an appearance, and proceeded to break up into piles of little disks, which separated and swam off as one watched them. The French poet, Catulle Mendés, imagined a world where the flowers flew about freely and the butterflies were fixed to stalks. His fancy is to some degree realized by the swimming away of the young jelly-fish from their stalks. There are a host of very minute jelly-fish, measuring when full grown only half an inch or less in diameter. They originate as buds from small branching polyps, one kind of which is common on oyster-shells, and is called "the herring-bone coralline." The dried skins of these coralline polyps (which are horny) are often to be picked up with masses of seaweed on the seashore after a storm. The little jelly-fish are the ripe individuals of the polyps, and produce eggs and sperm which grow to be polyp-trees. These, again, after growing and branching as polyps, give rise to little jelly-fish here and there on the tree, which in most kinds (though not in all) break off and swim away freely.

CHAPTER XII

SHRIMPS, CRABS, AND BARNACLES

WE have no word in English to indicate the varied crab-and-shrimp-like creatures of salt and fresh waters in the same way as "insect" designates the six-legged, usually winged, terrestrial creatures of many kinds—bettles, bees, bugs, two-winged flies, dragon-flies, day-flies, and butterflies. They are all "insects." Naturalists call the aquatic shrimp-and-crab creatures "crustaceans." Perhaps "crab" might be used in a large sense to include them all, together with the true crabs, as the Germans use their word, "krebs." The shore-crab is the most familiar of all crustaceans, in the living, moving condition. Boiled lobsters, prawns, and shrimps are more generally familiar members of the class, but the "undressed" living crab is better known to every one who has been on the seashore than the live lobster, prawn, and shrimp. Londoners have been heard to express interest in the curious blue variety of lobster caught on the coast, not being aware that the hot bath which he takes before he, too, is "dressed," causes his blue armour to change its colour to a brilliant scarlet. Occasionally a regular ordinary lobster is caught in which this change has occurred during life in the sea—and there are some enormous deep-sea prawns of a pound in weight which when living have a splendid crimson colour. A large series of "crustaceans," carefully prepared so as to show their natural colours in life, is exhibited in the Natural History Museum in Cromwell Road.

A curious kind of prawn (by name Althea rubra), of fair size, is found under "the low-tide rocks" in the Channel Islands, which not only is of a deep crimson colour, but snaps his fingers at you—or rather one of his fingers—or claws—when you try to catch him, making a loud crack audible at ten yards distance. The common big prawn, if you see him in a large vessel of sea-water with the light shining through him, appears very brilliantly marked with coloured bands and spots—reddish-brown, blue, and yellow—which are displayed on a transparent, almost colourless surface. Of course, boiling turns him pale red. A common smaller species of prawn when boiled is often sold as "pink shrimps," and lately a deep-sea prawn—a third species—has come from the Norwegian coast into the London market. There are many kinds which are not abundant enough to become "marketable." Prawns are at once distinguished from the true "brown shrimp" by having the front end of the body drawn out into a sharp-toothed spine, which is absent in the shrimp. Besides the prawns (Palæmon and Pandalus), the shrimp (Crangon), and the common lobster (Homarus), you may see in the London fish shops the large spiny lobster (Palinurus) called "langouste" by the French, and apparently preferred by them as a table delicacy to the common lobster, although it has no claws. It used to be called "craw-fish" or "sea craw-fish" in London; why, I am unable to say. The name was certainly bad, as it leads to confusion with the cray-fish, the fresh-water lobster of British and all European rivers (there are many other kinds of fresh-water lobsters in other parts of the world, as well as fresh-water prawns and crabs), whose English name is a curious corruption of the French one, "écrevisse" (cray-vees, cray-fish). Another lobster of our markets is the little one known as the "Dublin prawn," which is common enough on the Scotch and Norwegian coasts, as well as that of Ireland. Naturalists distinguish it as Nephrops Norvegicus. The great edible crab completes the list of British marketable crustaceans, but in Paris I have eaten, as well as at Barcelona, a very large Mediterranean prawn, three times as big as our biggest Isle of Wight prawns, but by no means so good. It is called "Barcelona prawn" and "Langostino" ("Penæus" by naturalists). In Madrid I have seen in the fish shops and eaten yet another crustacean—a very curious one—namely, a long-stalked rock-barnacle of the kind known to naturalists as Pollicipes.

That the barnacles—ship's barnacles (Fig. 10) and with them the little sea-acorns (Fig. 11), those terribly hard and sharp little white "pimples" which cover the rocks nearly everywhere just below high-tide mark, and have so cruelly lacerated the hands and shins of all of us who swim and have had to return to a rocky shore in a lively sea—should be included with crabs, lobsters, and shrimps as "crustaceans" must appear astonishing to every one who hears it for the first time. The extraordinarily ignorant, yet in their own estimation learned, fishermen of the Scottish coast will tell you with solemn assurance that the ubiquitous encrusting sea-acorns are the young of the limpet, whilst the creature living inside the shell of the long-stalked ship's barnacles has for ages been discoursed of by the learned as one of the marvels of the sea—nothing more or less than a young bird—the young, in fact, of a goose—the barnacle goose which, since it was thus proved to be a fish in origin, was allowed to be eaten by good Catholics on fast days! Two hundred years or more ago this story was discredited by serious naturalists, but the barnacles and sea-acorns were thought (even by the great Cuvier) to be of the nature of oysters, mussels, and clams (Molluscs), because of their possessing white hard shells in the form of "valves" and plates, which can open and shut like those of mussels. Their true history and nature were shown about eighty years ago by a great discoverer of new things concerning marine creatures, Dr. Vaughan Thompson, who was Army Medical Inspector at Cork, and studied these and other animals found in the waters of Queenstown Harbour.

The crab class, or Crustacea, have, like the insects, centipedes, spiders, and scorpions, a body built up of successive rings or segments. The earth-worms (as every one knows) and marine bristle-bearing worms also show this feature in the simplest and most obvious way. The vertebrates, with their series of vertebræ or backbone-pieces and the body muscles attached ring-wise to them, show the same condition. The marine worms have a soft skin and a pair of soft paddle-like legs upon each ring of the body, often to the number of a hundred such pairs. But the crab class and the classes called insects, centipedes, arachnids, and millipedes are remarkable for the hard, firm skin, or "cuticle," which is formed on the surface of their bodies and of their legs, which, as in the marine worms, are present—a pair to each body-ring or segment—often along the whole length of the body as in centipedes. This hard cuticle is impregnated with lime in the bigger members of the crab class, such as the lobster. It is not equally thick and hard all over the surface of the lobster, but is separated by narrow bands of thin, soft cuticle into a number of harder pieces, thus rendered capable of being bent or "flexed" on one another. Thus the body is jointed into a series of rings, and the legs are also divided each into several joints (as many as seven), which gives them flexibility and so usefulness of various kinds. The various joints are "worked" by powerful muscles, which are fixed internally to the cuticle and pass from one hard ring or segment, whether of body or of leg, to a neighbouring ring.

Every one knows the structure of a lobster's tail and of its legs, which can be readily examined in illustration of my statement, and the same structure can be seen in the leg of a beetle or a fly. Naturalists term all this series of creatures with hard-jointed cuticle, to which the muscles are attached, including the crab class, the insects, centipedes, spiders, and scorpions, "jointed-leg owners," or Arthropods. It is easy to appreciate this characteristic difference which separates the Arthropods from other animals. The sea-worms differ from them, in that they have soft cuticle, but stiffen and render their paddle-like legs firm by squeezing the liquid of the body into them in the same sort of way as the sea-anemones distend their tentacles with liquid, though in that case the liquid is sea-water taken in by the mouth. The Molluscs also distend their muscular lobe, or "foot" as it is called, by pressing the blood from the rest of the body into it, and so making it swell and become stiff, so that the muscles can work it; when not distended in that way it is flaccid. The Vertebrates (bony animals) and the star-fishes have again another and peculiar mechanism. Their muscles are attached to hard internal pieces, sometimes cartilaginous but often calcareous or bony, which are spoken of as "the internal skeleton." There are thus three distinct kinds of mechanism in animals for giving the necessary resisting surfaces, hinged or jointed to one another, and made to "play" one on the other by the alternate contraction and relaxation of the muscles attached to them.

The Arthropods differ among themselves in the number of body-rings, the enlargement or dwindling of certain rings, and the fusion of a larger or smaller number of the rings to form a composite head, or a jointless mid-body or hind-body. The successive legs are primarily and essentially like to one another, and each body-ring, with its pair of legs, is but a repetition of its fellows. At the same time, in the different classes included as "Arthropoda" a good deal of difference has been attained in the structure of the legs, and they have in each class a different form and character in successive regions of the body, distinctive of the class, and are sometimes, but not always, absent from many of the hinder rings. All these Arthropods agree in having a leg on each side immediately behind the mouth—belonging to a body-ring, which is fused with others to form the head—very specially shortened, of great strength and firmness, and shaped so as to be pulled by a powerful muscle attached to it, against its fellow of the opposite side, which is similarly pulled. These two stumpy legs form thus a powerful pair of nippers called "the mandibles." They are jaws, although they were in the ancestors of the Arthropods merely legs. These jaw-legs, or leg-jaws, are characteristic of all the crab class, as well as of the other Arthropods, but no bristle-worm or other animal has them. The jaws of marine worms are of a totally different nature. So are the jaws of snails, whelks, and cuttle-fish. Many of the crab class have not one only, but several, pairs of legs following the mouth converted into jaws. Thus, if you examine a big shore-crab, or, better, an edible crab, and a lobster, and a large prawn, you will find that they all have five pairs of legs converted into short foliaceous jaws (hence called "foot-jaws"), and overlying the first very strong pair, or mandibles.

Following these "foot-jaws" you find in a crab or a lobster the great nipping claws and the four large walking legs—the same in proportion and shape in crab, lobster, and prawn, much bigger than the foot-jaws. But the curious thing is that if you set them out and carefully compare them (for they are not simple jointed limbs, but each has two or even three diverging stems carried on a basal joint), you will find a strange and fascinating "likeness in unlikeness," or an agreement of the parts of which they are built, and yet a difference between all of them.

The rings of the body to which the jaw-legs and legs are attached are fused into one unjointed piece. The spine in front of the mouth and the support of the eyes and the feelers or "antennæ" are fused with that piece. It forms on the back a great shield—often called "the head"—which overhangs and is bent down over the sides of this region, so as to protect the gills, which you can see by cutting away the overhanging flap.

Following on the jaw-legs or foot-jaws and walking-legs, in the three crustaceans we are looking at, comes the jointed tail or hind-body, consisting of seven pieces. The first five rings of the tail have small Y-shaped legs, a pair to each ring. They are called "swimmerets," whilst the sixth has legs of the same shape, but very large and flat. In the middle between these large flat legs is the last ring, which has no legs, but is perforated by the opening of the intestine. You will see if you compare the crab and the lobster (or the prawn, which is very much like the lobster), that the crab has the so-called head (really head and mid-body combined) drawn out from side to side, so as to make it much wider than it is long. And, moreover, the jointed tail or

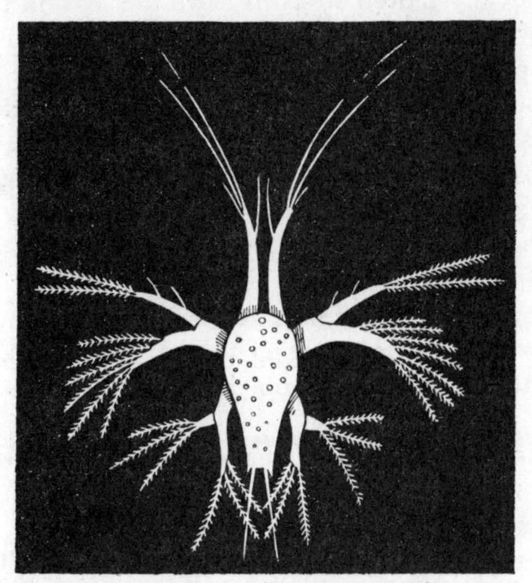

FIG. 9.—The larval or young form of Crustacea known as "the Nauplius." This is the "Nauplius" of a kind of Prawn. The three pairs of branched limbs are well seen. Much magnified.

hind-body seems at first sight to be absent in the crab. But if you turn the crab (a dead one) on his back, you will find that he has a complete tail, on the whole like that of the lobster, but pointed and bent forwards, and closely packed under the fused head and mid-body in a groove, from which you can raise it and turn it back.

We have not yet done with the various forms

assumed by the legs of our three crustaceans—for, actually in front of the mouth, there are two pairs of peculiarly altered legs. Originally in crab-ancestors, and at the present day in the very minute young stage of growth called "the Nauplius" (Fig. 9), the mouth was not behind these two front pairs. It has sunk back as it were, gradually moved so as to leave the legs in front of it. As we now see them in the crab, lobster, and prawn, the two pairs of legs in front of the mouth are jointed filamentous things—the feelers or antennæ—very long in prawns and lobsters, short in crabs. In the ancestors of crabs, lobsters, and prawns these feelers were undoubtedly swimming legs. In the "nauplius" stage (Fig. 9) of some prawns, and in many minute crustaceans often called "water-fleas," we find these feelers not acting as mere sensory organs of touch, but relatively strong and large, with powerful muscles, striking the water and making the little creatures bound or jump through it in jerks.

It has been discovered that in the growth from the egg of many crustaceans the young hatches out as a "nauplius" with only three pairs of legs. The front two pairs later gradually grow to be the feelers, the third pair become eventually the mandibles or first pair of jaw-legs. These legs all present themselves at first as active, powerful swimming "oars," beset with peculiar feathery hairs and not in the shape which they later acquire. The kite-shaped nauplius baby-phase, smaller than a small flea, with its three pairs of violently jerking legs, is a very important little beast. It is the existence of this young stage in the growth of barnacles and sea-acorns which has demonstrated that they are crustaceans, that is to say, belong to the crab class. The fixed shell-like barnacles and sea-acorns hatch from their eggs each as a perfect little "nauplius," like that drawn in Fig. 9. They swim about with jerking movements caused by the strokes of the two front legs and of the pair which will become the mandibles. Their limbs have the special form and are beset with the feather-like hairs, and the whole creature has the kite-like shape—characteristic of the nauplius young of other Crustacea. They are indeed indistinguishable from those young. Whilst it was the Army doctor, Vaughan Thompson, who discovered that barnacles are strangely altered "shrimps," it was Darwin who made one of the most interesting discoveries about them—a discovery of which he was always, and rightly, very proud—as I will explain in the next chapter.

CHAPTER XIII

BARNACLES AND OTHER CRUSTACEANS

THE ship's barnacle looks at first, when you see one of a group of them hanging from a piece of floating timber, like a little smooth, white bivalve shell, as big as your thumb-nail, at the end of a thickish, worm-like stalk, from one to ten inches long (Fig. 10). But you will soon see that there are not only two valves to the white shell, but three smaller ones as well as the two principal ones. This does not separate them altogether from the bivalve-shelled molluscs (mussels, clams, oysters), for the bivalve molluscs, which bore in stone and clay, have small extra shelly plates, besides the two chief ones, whilst the Teredo, or ship's worm—a true bivalve mollusc—has an enormously long, worm-like body which favours a comparison with it of the long-stalked barnacle. If a group of barnacles is floating attached to a piece of timber undisturbed in a tank of sea-water you will see the little shells gape, and from between them a bunch of curved, many-jointed feelers will issue and make a succession of grasping or clawing movements, as though trying to draw something into the shell, which, in fact, is what they are doing—namely industriously raking the water on the chance of bringing some particle of food to the mouth which lies within the shell (Fig. 10).

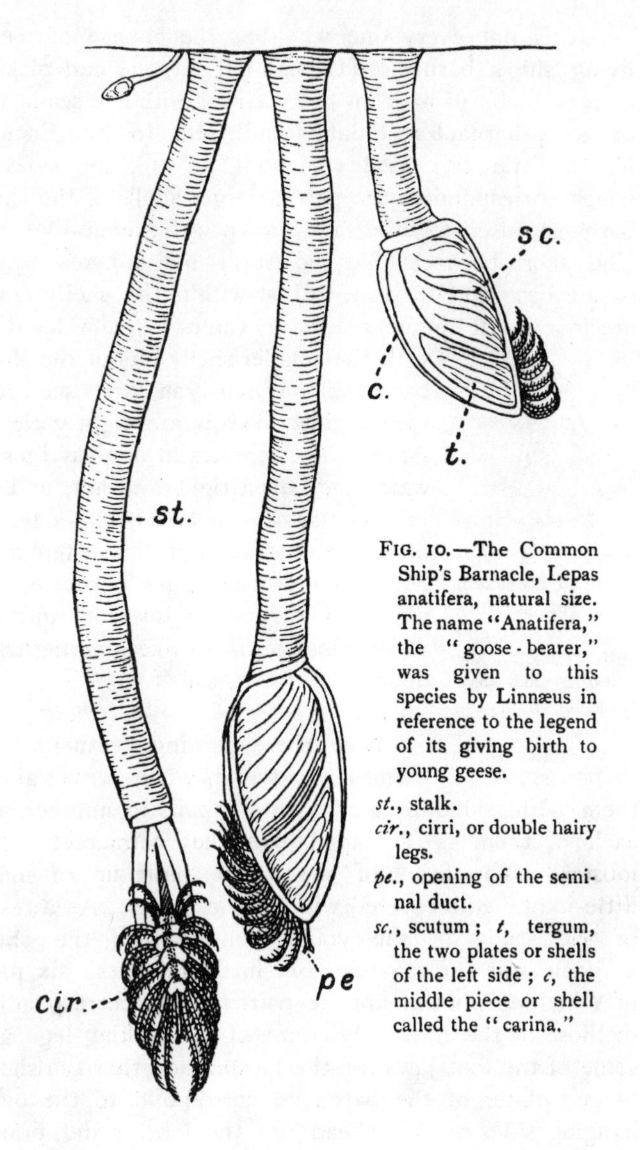

FIG. 10.—The Common Ship's Barnacle, Lepas anatifera, natural size. The name "Anatifera," the "goose-bearer," was given to this species by Linnæus in reference to the legend of its giving birth to young geese.

st., stalk.
cir., cirri, or double hairy legs.
pe., opening of the seminal duct.
sc., scutum; *t*, tergum, the two plates or shells of the left side; *c*, the middle piece or shell called the "carina."

It is not every one who has the chance of seeing living ship's barnacles (Lepas), but anyone can pick up a stone or bit of rock on the seashore with live sea-acorns or acorn-barnacles (Balanus) adherent to it. Each is like a little truncated volcano (Fig. 11), the sides of which correspond to the pair of larger shells of the ship's barnacle, fused together and grown into a cone-like wall. The acorn-barnacle has no stalk, but adheres by its broad base to the stone. Just within the shelly crater are four small hinged plates or valves in pairs, identical with the smaller shelly bits of the ship's barnacle. When you first see your specimen, the valves are tightly closed. After a few minutes in a glass of sea-water they open right and left, and up jumps—jack-in-the-box-wise—a tuft of bowing and scraping feelers or tentacles, like those of the ship's barnacle. If disturbed, they shoot inwards, and the valves close on them like a spring trap-door.

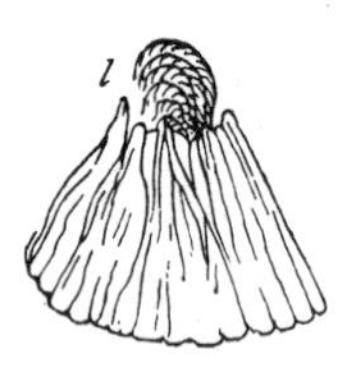

FIG. 11.—A large British Sea-acorn, Balanus porcatus, allied to the Ship's Barnacle. *l*, the feather-like legs issuing from the shell. Drawn of the natural size.

Now, these clawing, feathery little plumes are found, when we examine them with a hand-glass, to be six pairs in number, and each of them is Y-shaped, like the swimmerets of a lobster. The arms of the Y are built up of many little joints and covered with coarse hairs. As a result of the study of the young condition of the ship's barnacle and the sea-acorn, we find that these six pairs of Y-shaped plumes are six pairs of legs corresponding to those of the mid-body (some of the walking legs and some of the foot-jaws) of the lobster, and that the shelly hinged plates of the barnacles correspond to the over-hanging sides of the "head" of the lobster and prawn, which one can imagine to be hinged along a line running down the back so as to open like the covers of a book. There are very common little, free-swimming "water-fleas" (minute crustaceans) of many hundreds of kinds which have hinged shells of this description when in the full-grown condition, and it is found that the young barnacles and sea-acorns pass through a free-swimming phase of growth (the Cyprid stage), in which they greatly resemble these "water-fleas."

In fact, it is quite easy to hatch the young from the eggs of either ship's barnacles or acorn-barnacles at the right season of the year. They commence life as do so many Crustacea—in the "nauplius state," with three pairs of jerking limbs (Fig. 9). As they grow the overhanging pair of shells, delicate and transparent, appear; the three pairs of nauplius legs lose their swimming power; the most anterior (always called antennules in all crustaceans) become elongated and provided each with an adhesive sucker, on the face of which a large cement gland opens, secreting abundant adhesive cement; the second pair (antennæ) shrivel and disappear altogether; the third pair lose their long blades for striking the water and remain as simple, but strong, stumps—the mandibles! Two new pairs of little jaw-feet appear behind these, and farther back on the now enlarged body (the whole creature is not bigger than a small canary seed!) six pairs of Y-shaped legs appear and strike the water rhythmically, so that the little creature swims with some sobriety. The region to which these legs are attached is marked with rings or segments, and behind it follows a small, limbless, hind body of four segments, or joints, ending with two little hairy prongs like a pitchfork. The right and left movable, shell-like fold, or down-growth, of the sides of the body encloses the whole creature except the protruding antennules with their suckers.

In this condition it swims about for a time, and then, once for all, fixes itself by means of the suckers and their abundant cement, on to rock, stone, or floating wood—and there remains for the rest of its life (Fig. 12). It

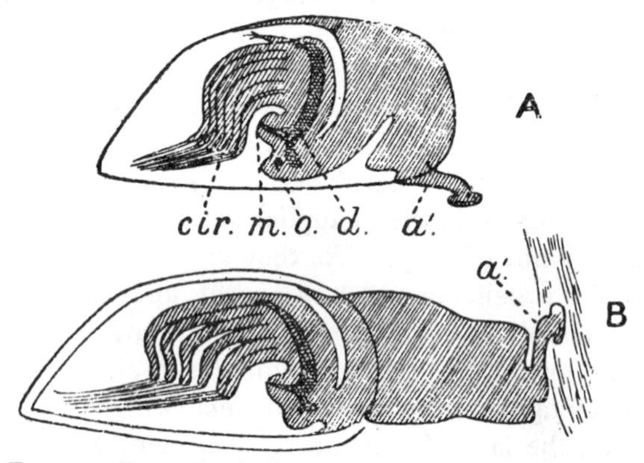

FIG. 12.—Two stages in the growth of the Common Barnacle from the Nauplius stage. Diagrammatic.

cir., the double legs or cirri; *m*, mouth; *o*, the single eye; *d*, the digestive canal.

a^1, one of the antennules or "feelers" (that of the right side of the head) provided with a sucking disk by means of which the young animal becomes fixed.

increases enormously in size, the delicate transparent shell develops into hard calcareous plates, opening and shutting on the hinge-line of the back. In the stalked kinds a peculiar elongated growth of an inch or several inches in length takes place between the mouth and the fixed suckers of the antennules (Figs. 10 and 12); in the short, so-called, "acorn" kinds, this stalk does not form, but a separate part of the shell grows into a ring-like protective wall or cone. The creature is thus actually fastened by its head—"upside down, with its legs sticking up" not in the air, but in the water. Those six pairs of Y-shaped legs, though no longer enabling the barnacle to swim, increase in relative size, and keep up their active movements. It is they which emerge like a plume when the valves of the shell open and carry on the rhythmic bowing and scraping movement described above.

The barnacles have, in fact, undergone a transformation which may be compared to that experienced by a man who should begin life as an active boy running about as others do, but be compelled suddenly by some strange spell or Arabian djin to become glued by the top of his head to the pavement, and to spend his time in kicking his food into his mouth with his legs. Such is the fate of the barnacles, and it is as strange and exceptional amongst crustaceans as it would be amongst men. Indeed, to "earn a living" human acrobats will submit to something very much like it. It is this change from the life of a free-living shrimp to that of a living lump, adherent by its head to rocks or floating logs, that Vaughan Thompson in 1830 discovered to be the story of every barnacle, and so showed that they were really good crustaceans gone wrong, and not molluscs. It is a curious fact that the young ascidian or sea-squirt which swims freely and has the shape of a tadpole, also when very young fixes itself by the top of its head to a rock or piece of seaweed, and remains immovable for the rest of its life. Though agreeing in their strange fixation by the head, the barnacle and the ascidian are very different kinds of animals. (For some account of the Ascidian the reader may consult the chapter "Tadpoles of the Sea" in "Science from an Easy Chair," Second Series. Methuen, 1912.)

The name "Cirripedes" is commonly used for the order or group formed by the barnacles—in allusion to the plume-like appearance of their "raking" legs. Stalked barnacles often are found in the ocean attached to floating pumice-stone, and one species has been discovered attached to the web of the foot of a sea-bird. They, like many other creatures, benefit by being carried far and wide by floating objects. Whales have very large and solid acorn-barnacles peculiar to them, fixed deeply in their skin. Others attach themselves to marine turtles.

With few exceptions the crustaceans are of separate sexes, male and female. But in nearly all classes of animals we find some kinds, even whole orders, in which the ovaries and spermaries are present in one and the same individual. "Monœcious" or "one-housed"—that is to say, possessing one house or individual for both ovaries and spermaries—is the proper word for this condition, but a usual term for it is "hermaphrodite." "Diœcious" is the term applied to animals or plants in which there are two kinds of individuals—one to carry the spermaries, the male, and the other to carry the ovaries, the female. It is probable that the monœcious condition has preceded the diœcious in all but unicellular animals. In vertebrate animals as high as the frogs and the toads we find rudimentary ovaries in the male, and in individual cases both ovaries and spermaries are well developed. Such a condition is not rare as an individual abnormality in fishes. In some common species of sea-perch (Serranus) and others it is not an exception but the rule.

Many groups of molluscs are monœcious, and it is not in any way astonishing to find a group of crustaceans which are so. The Cirripedes or barnacles are an example. It is probable that the presence of ovaries and spermaries in the same individual—the monœcious condition—is an advantage to immovable fixed animals. During the voyage of the "Beagle," and making use on his return of the collections then obtained, Darwin carried out a very thorough study of the Cirripedes of all kinds from all parts of the world. He worked out their anatomy minutely, classified the 300 different kinds then known, and described many new kinds. The stalked barnacles often occur in groups, the individuals being of different ages and sizes, the small young ones sometimes fixing themselves by their sucker-bearing heads to the stalks of their well-grown relatives. In all the varied kinds studied by Darwin he found that the full-grown individuals were monœcious—that is, of combined sex—as was known to be the case in those studied before his day. But Darwin made the remarkable discovery that in two kinds of stalked barnacles (not the common ship's barnacles), comprising several species, "dwarf males" were present perched upon the edge of the shell of the large monœcious (bi-sexual) individuals. These dwarf males were from one-tenth to one-twentieth the length of the large normal monœcious individuals, but usually possessed the characteristic details of the shell-valves and other features of the latter.

This existence of a sort of supernumerary diminutive kind of male as an accompaniment to a race of normal monœcious individuals was quite a new thing when Darwin discovered it. That all the males in some diœcious animals are minute as compared with the females was known, and has been established in the case of some parasitic crustaceans, in some of the wheel-animalcules, and in the most exaggerated degree in the curious worms, Bonellia and Hamingia. But the existence of "complemental males," as Darwin called them, existing apparently in order to fertilize the eggs should they escape fertilization by the ordinary monœcious individuals, was a new thing. And it was doubted and disputed when Darwin described his observations fifty-six years ago. They were, in fact, by many regarded as a distinct species parasitic upon the larger barnacles on which they were found until Darwin's conclusion as to their nature was confirmed by the report of Dr. Hoek, on the barnacles brought home by the "Challenger" expedition.

It is an interesting fact that recent studies have shown that in some of the barnacles with dwarf males (species of Scalpellum) the large individuals are no longer monœcious, but have become purely females, whilst in some other species dwarf males have been discovered which have rudimentary ovaries. Thus we get gradations leading from one extreme case to the other. Darwin always felt confidence in his original observations on this matter, and was proportionately delighted when, after thirty years, his early work was proved to be sound. In the Natural History Museum at the Darwin centenary in 1909, a temporary exhibition of specimens, note-books, and letters associated with Darwin's work, was brought together. His original specimens and drawings of Cirripedes and of the wonderful little "complemental males" of the barnacles were placed on view.

CHAPTER XIV

THE HISTORY OF THE BARNACLE AND THE GOOSE

THE curious belief, widely spread in former ages—that the creatures (described in the last chapter) called "barnacles" or "ship's barnacles"—often found attached in groups to pieces of floating timber in the sea as well as fixed to the bottoms of wooden ships—are the young of a particular kind of goose called "the barnacle goose," which is supposed to hatch out of the white shell of the long-stalked barnacle, is a very remarkable example of the persistence of a tradition which is entirely fanciful. It was current in Western Europe for six or seven centuries, and was discussed, refuted, and again attested by eminent authorities even as late as the foundation of the Royal Society—the first president of which, Sir Robert Moray, read a paper at one of the earliest meetings of the society in 1661, in which he described the bird-like creature which he had observed within the shell of the common ship's barnacle, and favoured the belief that a bird was really in this way produced by a metamorphosis of the barnacle.

The story was ridiculed and rejected by no less a philosopher than Roger Bacon in the thirteenth century, and was also discredited by the learned Aristotelian Albertus Magnus at about the same time. No trace of it is to be found in Aristotle or Herodotus or any classical author, nor in the "Physiologus." The legend seems to have originated in the East, for the earliest written statement which we have concerning it is by a certain Father Damien, in the eleventh century, who simply declares: "Birds can be produced by trees, as happens in the island of Thilon in India." We have also a reference to the same marvel in an ancient Oriental book (the "Zohar," the principal book of the Kaballah), as follows: "The Rabbi Abba saw a tree from the fruits of which birds were hatched." The earliest written statements of the legend are, it appears, to the effect that there is a tree which produces fruits from which birds are hatched. The belief in the story seems to have died out at the end of the seventeenth century, when the structure of the barnacle lying within its shell was examined without prejudice, and it was seen to have only the most remote resemblance to a bird. The plumose legs or "cirrhi" of the barnacle (Fig. 10) have a superficial resemblance to a young feather or possibly to the jointed toes of a young bird, and there the possibilities of comparison end.

The notion that a particular kind of black goose (a "brent"), which occurs on the marshy coast of Britain in great numbers, is *the* goose, *the* bird, produced by the barnacle was favoured by the fact that this goose does not breed in Britain, and yet suddenly appears in large flocks, in districts where barnacles attached to rotting timber are often drifted on to the shore. It was accordingly assumed by learned monks—*who already knew the traveller's tale*, that in distant lands birds are produced by the transformation of barnacles—that this goose is the actual bird which is bred from the barnacles, and it was accordingly called "the barnacle goose." I think that this identification was due to the exercise of a little authority on the part of the clergy in both France and Britain, who were thus enabled to claim the abundant "barnacle goose" as a fish in its nature and origin rather than a fowl, and so to use it as food on the fast-days of the Church. Pope Innocent III (to whom the matter was referred) considered it necessary in 1215 to prohibit the eating of "barnacle geese" in Lent, since although he admitted that they are not generated in the ordinary way, he yet maintained (very reasonably) that they live and feed like ducks, and cannot be regarded as differing in nature from other birds.

Thus we see that in early and even later days a good deal hung on the truth of this story of the generation of barnacle geese. The story was popularly discussed by the devout and by sceptics, and appears to have been known in France as "l'histoire du canard." At last in the seventeenth century it was finally discredited, owing to the account given by some Dutch explorers of the eggs and young of the barnacle goose—like those of any other goose—and its breeding-place in the far north on the coast of Greenland. The discredited and hoary legend now became the type and exemplar of a marvellous story which is destitute of foundation, and so the term "un canard" (short for histoire d'un canard), commonly applied in French to such stories, receives its explanation. Our own term for such stories, in use as long since as 1640, namely, "a cock-and-bull story," has not been traced to its historical source.[1]

That the story of the goose or duck and the

[1] Probably it means "a silly story told by a cock to a bull!" as suggested by the French word *coq-à-l'âne*, which means a story told or fit to be told by a cock to an ass!

transformed barnacle was a popular one in Shakespear's time, whether believed or disbelieved, appears from his reference to barnacles in "The Tempest." Caliban says to Stephano and Trinculo, when they have all three been plagued by Prospero's magic, and plunged by Ariel into "the filthy mantled pool" near at hand, "dancing up to their chins": "We shall lose our time and all be turned to barnacles, or to apes with foreheads villainous low." Probably enough, this is an allusion to the supposed Protean nature of barnacles. They are not alluded to elsewhere in Shakespear.

One of the most precise accounts of the generation of geese by barnacles is that of the mediaeval historian Giraldus Cambrensis, who visited Ireland and wrote an account of what he saw in the time of Henry II, at the end of the twelfth century. He says: "There are in this place many birds which are called Bernacæ; Nature produces them, against Nature, in a most extraordinary way. They are like marsh-geese, but somewhat smaller. They are produced from fir timber tossed along the sea, and are at first like gum. Afterwards they hang down by their beaks as if they were a seaweed attached to the timber, and are surrounded by shells in order to grow more freely. Having thus in process of time been clothed with a strong coat of feathers, they either fall into the water or fly freely away into the air." "I have frequently seen," he proceeds, "with my own eyes, more than a thousand of these small bodies of birds, hanging down on the seashore from a piece of timber, enclosed in their shells and ready formed. They do not breed and lay eggs like other birds; nor do they ever hatch any eggs nor build nests anywhere. Hence bishops and clergymen in some parts of Ireland do not scruple to dine off these birds at the time of fasting, because they are not flesh nor born of flesh!"

It is noteworthy that Giraldus does not state—in accordance with the tradition as reported by earlier writers—that there is a tree the buds of which become transformed into the geese, but says merely that the "small bodies of birds," clearly indicating by his description groups of ship's barnacles, are "produced from fir timber tossed along the sea." It is also noteworthy that he calls the geese themselves "Bernacæ," which is the Celtic name for a shell-fish.

Later the belief seems to have reverted to the older tradition, or probably enough the complete story, including the existence of the bird-producing tree, existed in its original form in "seats of learning" in other parts of the British Islands outside Ireland, and also in Paris and other places in Western Europe. For we find that in 1435 the learned Sylvius, who afterwards became Pope Pius II, visited King James of Scotland in order, among other things, to see the wonderful tree which he had heard of as growing in Scotland from the fruit of which geese are born. He complains that "miracles will always flee further and further," for when he had now arrived in Scotland and asked to see the tree, he was told that it did not grow there, but farther north, in the Orkneys. And so he did not see the tree.

In 1597, John Gerard, in the third book of his "Herbal, or History of Plants," writes as follows: "There are found in the north parts of Scotland and the Islands adjacent called Orchades, certaine trees whereon do grow certaine shell-fishes of a white colour tending to russett, wherein are contained little creatures which shels in time of maturity doe open and out of them grow those little living things which, falling into the water, doe become foules whom we call Barnacles, in the north of England Brent Geese, and in Lancashire Tree Geese." Gerard is here either adopting or suggesting an identification of the tradition of the tree which produces birds from its buds, with the floating timber bearing ship's barnacles, which were supposed to give birth to the brent geese. He does not say that he has seen, or knows persons who have seen, the barnacles attached to the branches of living trees. Nevertheless, he gives a picture of them so attached (Fig. 13). It has been suggested, in later times, that such a fixation of barnacles to the branches of living trees might occur in some of the sea-water lochs of the west of Scotland,—just as oysters become attached to the mangrove trees in the West Indies,—and it has further been suggested that willows might thus droop their branches into the sea-water, and that the catkins on the willow-shoots might be taken for an early stage of growth of the barnacles; but I have not come across any record of such fixation of barnacles on living shrubs or branches of trees, and I am inclined to think that Gerard's story of what occurs in the distant Orkneys is merely an attempt to substantiate the bird-producing tree of the Oriental story, by quietly assuming that the sea-borne timber covered with barnacles existed somewhere as living trees and exhibited this same property of budding forth barnacles which on opening liberated each a minute gosling. Gerard continues as follows: "But what our eyes have seen and hands have handled we shall declare." There is, he tells us, a small island in Lancashire called the Pile of Foulders, and there rotten trees and the broken timbers of derelict ships are thrown up by the sea. On them forms "a certain spume or froth which in time breeds into certaine shells." He

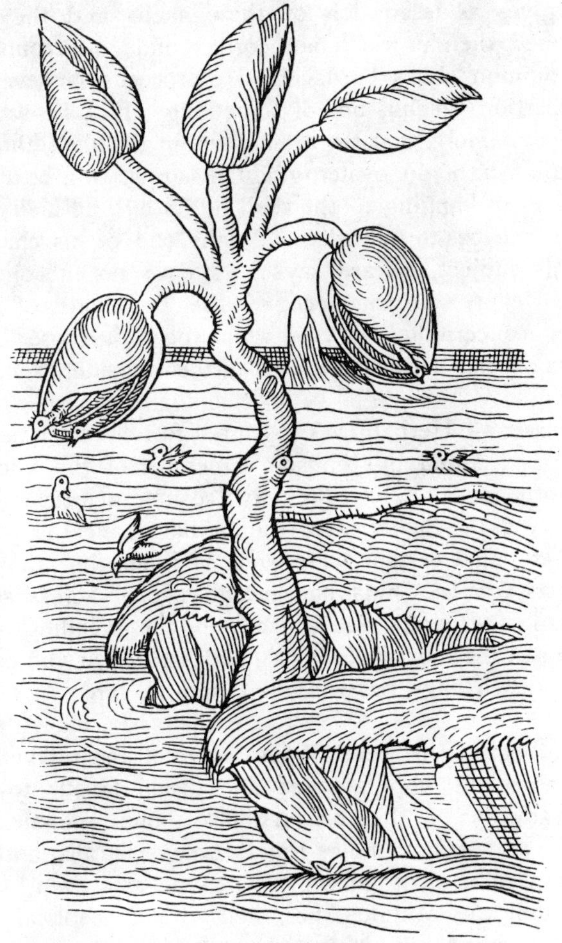

FIG. 13.—The picture of the "Goose Tree," copied from the first edition of Gerard's "Herbal."

The fruit-like oval bodies are "barnacles" (Lepas) fancifully represented as growing like buds or fruit on a little tree. Some of the young geese are drawn as in the act of escaping from the barnacle-shells, and others are represented swimming in the water.

then gives a description of these shells and the fish contained therein, which is a correct enough account of the common ship's barnacle. He proceeds, however, to an assertion which is not of something which he saw or handled, namely, that the animal within the shell, though like the fish of an oyster, gradually grows to a bird and comes forth hanging to the shell by its bill. Finally, he says, it escapes to maturity. At the end of his chapter on this subject, Gerard says: "I dare not absolutely avouch every circumstance of the first part of this history concerning the tree which beareth those buds aforesaide, but will leave it to a further consideration."

Gerard's "Herbal" was reprinted forty years later (in 1636) and edited by Johnson, a member of the Society of Apothecaries. He writes with contempt of Gerard's credulity as to the story of the barnacle and the goose, and states that certain "Hollanders" in seeking a north-east passage to China had recently come across some islands in the Arctic Sea which were the breeding-place of the so-called barnacle goose, and had taken and eaten sixty of their eggs, besides young and old birds.

Probably there were always lovers of the marvellous and the occult who favoured and would favour to-day the tradition of the conversion of one animal into another and such wonders; and there were also both in the days of ancient Greece and Rome, and even in the darkest of the Middle Ages, men with a sceptical and inquiring spirit, who accepted no traditional testimony, but demanded, as the basis of their admitting something unlikely as nevertheless true, the trial of experiment and the examination of specimens. What has happened since Gerard's time and the incorporation of the Royal Society in 1662, is that the sceptical men have got the upper hand, though not without much opposition. In this country, owing to the defective education administered in our public schools and older universities, there is still quite a large number of well-to-do people ready to believe in any "occult" imposture or fantasy that may be skilfully brought to their notice.

On the other hand, we must bear in mind when we consider these strange beliefs held by really learned and intelligent men in the past, that the investigation of nature had not advanced very far in their time. It was not held, as it is to-day, as an established fact that living things are generated only by slips or cuttings of a parent or from eggs or germs which are special detached particles of the parent. It was held to be a matter of common observation and certainty that all sorts of living things are "spontaneously generated" by slime, by sea foam, by mud, and by decomposing dead bodies of animals and trees. It was also held, in consequence of a blind belief in, and often a complete misunderstanding of, the legends and fairy tales of the ancients and of the preposterous "Bestiaries" and books on magic which were the fashion in mediaeval times, that it is quite a usual and natural thing for one animal or plant to change into another. Hence there was nothing very surprising (though worthy of record) in a barnacle changing into a young goose, or in the buds of a tree becoming in some conditions changed into barnacles!

So, too, the notion that rotting timber can "generate" barnacles was not, to our forefathers, at all out of the way or preposterous. Sir Thomas Browne in 1646 was unable to make up his mind on this matter, and believed in the spontaneous generation of mice by wheat, to which he briefly alludes in his curious book called

"Pseudodoxia Epidemica, or an Enquiry into Vulgar and Common Errors." The account of the creation given by the poet Milton was based upon the belief in the daily occurrence of such spontaneous generation of living things of high complexity of structure and large size, from slime and mud. The process of creation of living things conceived by him was but a general and initial exhibition of an activity of earth and sea which in his belief was still in daily operation in remote and undisturbed localities.

In 1668 the Italian naturalist, Redi, demonstrated that putrefying flesh does not "spontaneously breed" maggots. He showed that if a piece of flesh is protected by a wire network cover from the access of flies, no maggots appear in it, and that the flies attracted by the smell of the meat lay their eggs on the wire network, unable to reach the meat, whilst if the wire cover is removed they lay their eggs on the meat, and from them the maggots are hatched. It took a long time for this demonstration by Redi to affect popular belief, and there are still country folk who believe in the spontaneous generation of maggots.[1]

But few, if any, persons of ordinary intelligence or education now believe that these sudden productions of living things, without regular and known parentage, take place. The spontaneous generation of large, tangible creatures having ceased to be an article of general belief, the conviction nevertheless persisted for some time that at any rate minute microscopic living things were generated without parentage. This theory was more difficult to test on account of the need for employing

[1] See the chapter, "Primitive Beliefs about Fatherless Progeny," in "Science from an Easy Chair," Second Series.

the microscope in the inquiry, which was not brought to a high state of efficiency until the last century. By experiments similar to those of Redi, it was shown in the first half of last century by Theodor Schwann that even the minute bacteria do not appear in putrescible material when those already in it are killed by boiling that material, and when the subsequent access to it of other bacteria is prevented by closing all possible entrance of air-borne particles, or insect carriers of germs. It took another fifty years to thoroughly establish by observation and experiment the truth of Schwann's refutation of the supposed "spontaneous generation" of the minutest forms of life.

As an example of the strange incapacity for making correct observation and the failure to record correctly things observed which are frequently exhibited by the most highly placed "men of education," as well as by uneducated peasants and fisher folk, we have the short paper entitled, "A Relation concerning Barnacles," by Sir Robert Moray—the first president of the Royal Society of London (from 1661 until its incorporation in 1662)—a very distinguished man, and an intimate friend of King Charles II. This paper was read to the society in 1661 and published in 1677 in vol. xii. of the "Philosophical Transactions." Sir Robert relates how he found on the coast a quantity of dead barnacles attached to a piece of timber, and that in each barnacle's shell was a bird. He writes: "This bird in every shell that I opened, as well the least as the biggest, I found so curiously and completely formed that there appeared nothing wanting, as to the external parts, for making up a perfect sea-fowl; every little part appearing so distinctly that the whole looked like a large bird seen through a concave or diminishing glass, colour and feature being everywhere so clear and near. The little bill like that of a goose, the eyes marked, the head, neck, breast, wings, tail and feet formed, the feathers everywhere perfectly shaped and blackish coloured, and the feet like those of other waterfowl—*to my best remembrance.* All being dead and dry, I did not look after the inward parts of them." If the reader will now look at Fig. 15, C, which represents the soft parts of a barnacle when the shells of one side are removed, he will see how far Sir Robert Moray must have been the victim—as so many people naturally are under such circumstances—of imagination and defective memory when he wrote this account. I have put into italics in the above quotation from his "Relation" his confession that he is writing, not with his specimens before him, but from remembrance of them. Moreover, he tells us, with admirable candour, that the specimens were dead and dry when he examined them! One could not desire a better justification for the motto adopted by the Royal Society, "Nullius in verba," and for the procedure upon which in its early days the Society insisted—namely, that at its meetings the members should "bring in" a specimen or an experiment, and not occupy time by mere relations and reports of marvels. It is necessary even at the present day to insist on such demonstration by those who urge us to accept as true their relations of mysterious experiences with ghosts, and their "conviction" that they have conversed with "discarnate intelligences."

CHAPTER XV

MORE AS TO THE BARNACLE AND THE GOOSE

IT is clear that there was a widespread tradition known to the learned in the early centuries of the Christian era, according to which there existed in some distant Eastern land a tree which bore buds or fruits which became converted into birds. Connected with this, and perhaps really a part of it, there existed a tradition that marine "barnacles" gave birth to geese from within their shells, or are in some way converted into geese. The two stories were in some localities and narrations combined, though in others they were distinct. On the coast of Ireland the early missionaries of the Church (learned men acquainted with the traditions of their time) identified the migratory brent goose with the bird said to be produced by the barnacle; and elsewhere, on the Scottish coast, the barnacles were (it was reported) found growing on trees. There is no such resemblance between barnacles and brent geese as to have suggested to the Irish monks the regular and natural conversion of one into the other. It seems most probable that the learned churchmen knew the traditional story already before arriving in Ireland, and applied it to the barnacles and the geese which they discovered around them. Eventually the word "barnacle" without qualification was applied to the geese, as we see in

Gerard's account given in the last chapter. Is there, it may be asked, anything further known as to such a tradition, and the place and manner of its origin? In the absence of such knowledge, an ingenious attempt was made by my old friend, Professor Max Müller, to account for the tradition by the similarity of the names, which he erroneously supposed had been given *independently* to the barnacle and to the "Hibernian" goose. I will refer to this below, but now I will proceed to give the most probable solution of the mystery as to the tradition of the tree, the goose, and the barnacle. Its discovery is not more than twenty years old, and is due to M. Frederic Houssay, a distinguished French zoologist of the Ecole Normale, who published it in the "Revue Archeologique" in 1895. It has not hitherto been brought to the notice of English readers, and I shall therefore give a full account of it.

The solution is as follows: The Mykenæan population of the islands of Cyprus and Crete, in the period 800 to 1000 years before Christ, were great makers of pottery, and painted large earthernware basins and vases with a variety of decorative representations of marine life, of fishes, butterflies, birds, and trees. Some of these are to be seen in the British Museum at Bloomsbury, where I examined them a few years ago. Others have been figured by the well-known archæologists, MM. Perrot and Chipiez, in the sixth volume of their work, "L'Ossuaire de Crète." M. Perrot consulted M. Houssay, in his capacity of zoologist, in regard to these Mykenæan drawings, which bear, as M. Houssay states, the evidence of having been designed *after nature* by one who knew the things in life, although they are not slavishly "copied" from nature. These early Mykenæan painters on pottery were members of a community who worshipped the great mother—" Nature "—as Astarte or Aphrodite risen from the foam of the sea. Being sailors and fishermen, marine life was even more familiar to them than that of the land, and they placed little models of

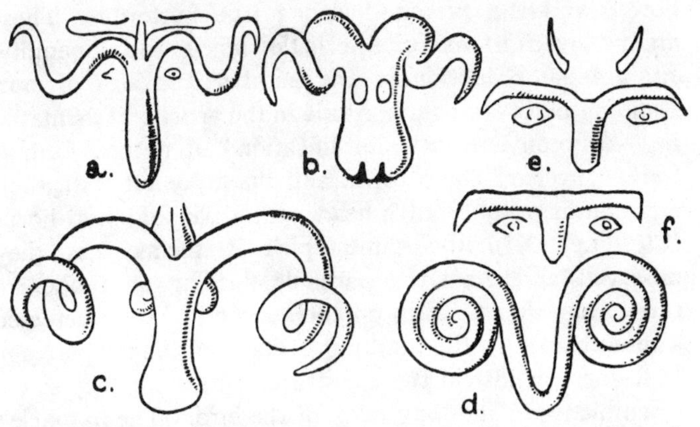

FIG. 14.—Fanciful designs by Mykenæan artists, showing change of the cuttle-fish (octopus or "poulpe") into a bull's head and other shapes.

a, Octopus drawn on a goblet from Crete, the arms reduced to two, the eyes detached.

b and c, Bull's head variations of the octopus, from designs found at Koban in the Caucasus.

d, Spiral treatment of the arms of the octopus (a pose actually seen in living specimens).

e, f, Human faces painted on Cretan jars across the whole width of the neck, the design being derived from the octopus with detached eyes as in Fig. a. Such designs survive long after their origin is forgotten, as (according to M. Houssay) the legend of the barnacle and the goose survived two thousand years after the Mykenæan drawings assimilating one to the other had been forgotten.

sea animals as votive offerings in the temples of the great mother, and also honoured her in decorating their pottery with marine creatures. The little fish, Hippocampus, called the sea-horse, the sea-urchin, the octopus, the argonaut and its floating cradle, the sea-anemone, and the butterfly-like Pteropod, were subjects used by these artists for which they found terrestrial counterparts. The sea-horse was convertible decoratively into a true horse, with intermediate phases imagined by the artists; the sea-urchin into a hedgehog, the sea-anemone into a flower, and the Pteropod into a true butterfly. These artists loved to exercise a little fancy and ingenuity. By gradual reduction in the number and size of outstanding parts—a common rule in the artistic "schematizing" or "conventional simplification" of natural form—they converted the octopus and the argonaut, with their eight arms, into a bull's head with a pair of spiral horns (Fig. 14). In the same spirit it seems that they observed and drew the barnacle floating on timber or thrown up after a storm on their shores. They detected a resemblance in the marking of its shells to the plumage of a goose, whilst in the curvature of its stalk they saw a resemblance to the long neck of the bird. The barnacle's jointed plumose legs or cirri and other details suggested points of agreement with the feathers of the bird. They brought the barnacle and the goose together, not guided thereto by any pre-existing legend, but by a simple and not uncommon artistic desire to follow up a superficial suggestion of similarity and to conceive of intermediate connecting forms. Some of their fanciful drawings with this purpose are shown in Figs. 15, 16, and 17. These (excepting the drawing of the barnacle lying within its opened shell) are copied from M. Houssay's paper on the subject, and were taken from the work of M. Perrot on Cretan pottery.

The intention of the artist to fantastically insist on intermediate phases between goose and barnacle is placed beyond doubt by certain details. For instance, in Fig. 16, the little jointed processes on the back

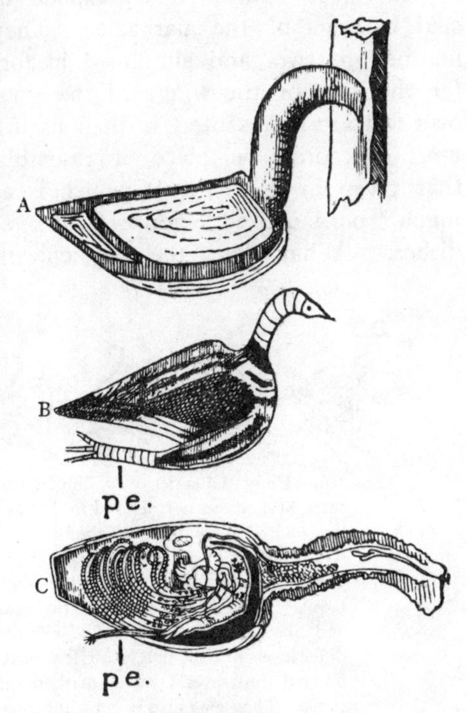

FIG. 15.—The Goose and the Barnacle.

A, Drawing of a Ship's Barnacle attached to a piece of timber by its "peduncle" or stalk, which represents the neck of a goose, if we regard the shell-covered region as the goose's body. From a sketch by M. Frederic Houssay published in the "Revue Archæologique," January 1895.

B, Copy of a drawing on an ancient Mykenæan pot found in Crete, and figured by M. Perrot in his "Ossuaire de Crète" vol. vi. p. 936. It is a fantastic blend of the goose and the barnacle. The barnacle's stalk is given a beak and an eye; the body of the bird corresponds to the shells of the barnacle both in shape and marking. There are no wings or legs, but the curious single limb which I have marked pe is obviously the same thing as that marked pe in figure C, which represents the barnacle when cut open so as to show the structures within the shell. pe is the rod-like body at the end of which the seminal duct opens. It is seen in the drawing of the expanded barnacle (Fig. 10), lying between the two groups of six forked and jointed legs or "cirri."

C, A correct modern drawing of a ship's barnacle, with the shells of one side removed so as to show the six double legs of one side, the seminal rod (pe), and the internal organs. This is what Sir Robert Moray and his mediaeval predecessors saw on opening the barnacle's shell and described as "a young bird complete in every detail."

of the goose marked a, correspond in position to th
cirri or legs of the barnacle. They are reduced i
number to two, and simplified in form so as to pas
for the tips of the wings of the goose. The goose'
own feet are represented in their natural position. Th
most extraordinary piece of resemblance in detail i
that given in Fig. 15, B, which is a copy of a ver
much "barnaculized" goose from one of these ancien
dishes. What does the Mykenæan artist mean t

FIG. 16.—Copy of a series of modified geese painted on an early Mykenæan pot, figured by M. Perrot. Each has two jointed appendages on the back, which suggest the wing feathers of the bird or two of the jointed legs (cirri) of the barnacle, which issue in life from this part of the barnacle's shell. The legs of the geese are very small and absent in the fifth. The markings on the body differ in each bird, but recall the shell of the barnacle divided into several valves marked with parallel striations. They may also pass for the plumage of the bird.

represent by the strange single leg-like limb marked pe? When we carefully examine the barnacle's soft body concealed by its shell, it becomes obvious that this leg-like thing corresponds to the single stalk-like body, ending in a bunch of a few hairs which is marked pe in Fig. 15, C. This last-named figure is a careful modern representation of the soft living barnacle, as seen when the shells of one side are removed. The cylindrical body pe of Fig. 15, C, which is drawn by the Mykenæan artist on an exaggerated scale in Fig. 15, B, is the external opening of the semina duct of the barnacle. It is remarkable that the Mykenæan pottery-painter had observed the soft "fish" of the barnacle so minutely as to select this unpaired and very peculiar-looking structure, and represent it of exaggerated size attached in its proper position on the barnacle-like body of a goose. This very striking transference of a peculiar and characteristic organ of the barnacle to the body of the goose by the artist seems not to have been noticed by M. Houssay.

FIG. 17.—Two drawings on pottery of modified geese, from Perrot's "Ossuaire de Crète." The three lines above the back of the upper figure probably represent the legs or cirri of the barnacle, which are represented by two jointed appendages in the geese shown in Fig. 16.

M. Houssay further points out the existence on some of the Mykenæan pottery of drawings (see "L'Ossuaire de Crète," by MM. Perrot and Chipiez) of leaves attached to tree-like stems. These leaves (Fig. 18, a, b, c) exhibit the same markings ("venation") which we see on the bodies of the geese in Fig. 16, especially the middle one of the five. The leaves (or fruits?) copied by M. Houssay from the Mykenæan pottery are attached in a series to a stem — but no one, at present, has suggested what plant it is which is represented. The corners of the leaf or fruit to the right and left of its stalk are thrown into a spiral—and the half leaf or half fruit represented in Fig. 18, b, leads us on to that drawn in Fig. 18, c, in which the spiral corner is slightly modified in curvature so as to resemble the head and neck of the goose as drawn in Fig. 16. Though Fig. 18, c, is as yet devoid of legs or wing feathers (compare Fig. 16, d), the black

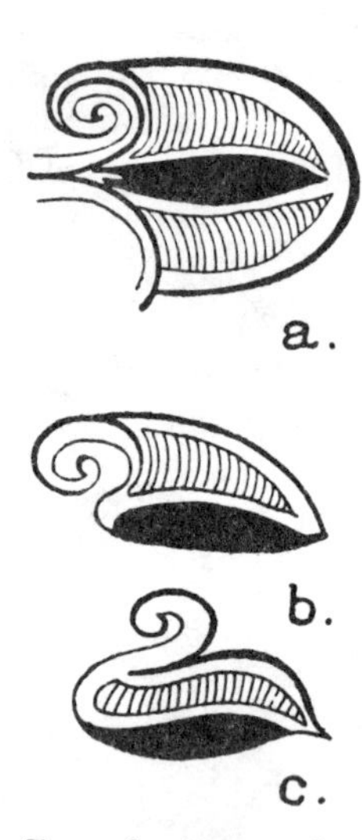

FIG. 18.—Leaves from the tree, drawn on a Mykenæan pot which, according to M. Perrot, are fancifully designed so as to assume step by step (a, b, c) the form of a goose. This appears either to represent the tree which, according to legend, produced birds as buds on its branches, or to be a fanciful design which gave rise to that legend. The artist's intention of making the leaf gradually pass into the semblance of a goose, is strongly emphasized by the purely fanciful "venation" of the leaf which agrees with the equally fanciful ornament of the bodies of the geese in Fig. 16, especially the middle one of the series.

band along the belly with the band of vertical markings above it agrees closely with the design on the body of the middle goose of the series drawn in Fig. 16. As these are associated in the decoration of the Mykenæan artists, it is fairly evident that the intention has been to manipulate the drawing of the leaf or fruit so as to make it resemble the drawing of the goose, whilst that in its turn is modified so as to emphasize or idealize its points of resemblance to a barnacle.

It is true enough that the drawings from Mykenæan pots here submitted cannot be considered as a complete demonstration that the legend of the tree-goose originated with these drawings. But it must be remembered that we have only a small number of examples of this pottery surviving from a thousand years B.C. It is probable that the fanciful decorative design of a master artist was copied and used in the painting of hundreds of pots by mere workmen or inferior craftsmen, and that more complete and impressive designs showing the fanciful transformation of leaf or fruit to goose, and of goose to barnacle, existed both before and after the making of the particular pots and jars which have come down to us. The supposition made by M. Houssay (which I entirely support) is that some later Levantine people—to whom these decorated pots or copies of their decorations became known either in the regular way of trade or as sailors' "curios"—were led to attempt an explanation of the significance of the pictures drawn upon them, and in accordance with a well-known and rooted tendency—interpreted the fancies of the artist as careful representations of astonishing fact. The existence of a tree which produces buds which become birds, and of a barnacle which becomes transformed into a goose—is the matter-of-fact interpretation of the few pictures of these animals which have come down to us, modern men, painted on the few pots of that remote Mykenæan industry now in our museums. It is not at all unlikely that in the vast period of time between 1000 B.C. and 1000 A.D., the more striking of these designs had been copied and familiarized in some part of the ancient world. It is true that we do not at present know in what part: we have not yet come across these designs of later date than 800 B.C. The absence of the story of the tree-goose from Greek and Roman lore is striking. Neither Aristotle nor Herodotus knew of it, although it has been erroneously stated that they refer to it. Yet the source of it was there in the Greek isles almost under their noses (if one may speak of the noses of such splendid and worshipful men of old) in the artistic work—otherwise not unknown to the Greeks—of a civilization which preceded their own by hundreds of years. There is other and ample evidence—as for instance that of the representation of the "flying gallop" (see "Science from an Easy Chair," Second Series, pp. 57 and 63), showing that Mykenæan art had little or no direct effect on the Hellenes, although the reputation of the skill of the old race in metal work came through many generations to them. Mykenæan art seems to have migrated with Mykenæan settlers to the remote region of the Caucasus. In the necropolis of Koban and other remote settlements, Mykenæan designs in bronze and gold—including the horse in flying gallop and octopods transformed to bull's heads—have been found and pictured (Ernest Chantre, "Recherches anthropologique dans Caucase," 4 vols.: Paris, 1886). They are believed to date from 500 B.C. It is possible that in such remote regions or in some of the Greek islands the pictures of the tree-goose and the barnacle may have survived until the new dispensation — that is, until the days of the Byzantine Empire. Once we can trace either the pictures or the legend up to that point, there is no difficulty about admitting the radiation of the wonderful story from that centre to the Jews of the Kabbalah, to Arabic writers, and so to the learned men of the Christian Church and the seats of learning throughout Europe and a great part of Asia.

Of the history of the legend during two thousand years we have no actual knowledge. It remains for investigation. But undoubtedly these Mykenæan pottery paintings remove the origin of the story to a period two thousand years older than that of the Irish monks.

One additional fact I may mention as to the existence of the goose and barnacle legend in the East. I am informed that in Java there is, according to "native" story, a shell-fish the animal of which becomes transformed into a bird—said to be a kind of snipe—and flies from the shell. I have been shown the shell by a Dutch lady who has lived in Java. It is a large fresh-water mussel, one of the Unionidæ. I have failed to obtain, after inquiry, any further information as to the prevalence or origin of this story in Java, and hope that some one who reads this page may be able to help me.

Before leaving the story of the goose and the barnacle, the explanation of the myth given by Prof. Max Müller in his lectures on the science of language nearly fifty years ago, should be cited. It is an excellent example of the misuse of hypothesis in investigation, and the attempt to explain something which we cannot get at and examine by making a supposition which it is even more difficult to examine and test.

Max Müller made use of the observation—a perfectly true and interesting one—that a whole people or folk will be led to a wrong conclusion, or to a belief in some strange and marvellous occurrence, by the misunderstanding of a single word, attributing to that word a sense which now fits the sound, but one quite different from that with which the word was originally used in the tradition or history concerned. Words are, in fact, misinterpreted after a lapse of time, or when imported from distant lands, just as we have seen that pictures and sculpture often have been. For instance, Richard Whittington, who was Lord Mayor of London in 1398 and other later years, did business in French goods, which was spoken of in the city as "achat," and pronounced "akat." Hence in later centuries, when the prevalence of Norman French was forgotten, it was stated (in a play produced in 1605) that Whittington owed his fortune to "a cat," and the story of the wonderful cat and its deeds was built up "line upon line" or "lie upon lie." Max Müller suggested that the story of the barnacle and the goose could be similarly explained. The brant or brent goose which frequents the Irish shore was, he supposes, called "berniculus" by the Latin-speaking clergy as a diminutive of Hibernicus, meaning "Irish." There is absolutely no evidence to support this. Max Müller supposes that Hibernicus became "Hiberniculus," and then dropping the first syllable became "Berniculus," and that this word was applied to the "Irish goose." It might have been, but there is nothing to show that it was. Meanwhile the ship's barnacle and other sea-shells were called in the Celtic tongue "barnagh," "berniche," or "bernak," and the hermit-crab is still called on the Breton coast, "Bernard l'hermite," a modification of "bernak l'hermite." There is no doubt that the word "barnacle" as applied to the stalked shell-fish growing on ships' bottoms is a diminutive of the Celtic word "bernak," or "barnak." It became in Latin "barnacus," and then the diminutive "barnaculus," and so "barnacle" was used for the little stalked shell fish encrusting old timber. According to Max Müller later generations thus found the two animals, goose and shell-fish, called by the same name, "bernikle," or "barnacle." "Why?" they would ask: and then (he supposes) they would compare the two and detect points of resemblance, until at last a very devout and astute monk had the happy thought of declaring that the Hibernian goose was called "berniculus," or "barnak-goose," *because* it did not breed from eggs as other birds do, but is hatched out of the shell of the shell-fish, also very naturally and rightly called "berniculus," or barnak, as any one may see by carefully examining the fish contained in the shell of the barnacle or little stalked "barnak," which has the complete form of a bird. Since, however, it is not a bird, but a fish in nature and origin, this holy man declared that the "berniculus," or "barnacle-goose," may be eaten on fast days. Max Müller's explanation of the origin of the story is too adventurous in its unsupported assumption that the particular goose associated with the story was peculiarly Irish, or that, in fact, any kind of goose was so. He also put aside the evidence of Father Damien (earlier than the Irish story of Giraldus) referring the goose-tree to an island in the Indies, and the report cited in the Oriental book the "Zohar." However plausible Max Müller's theory may have appeared, it absolutely crumbles and disappears in the presence of the Mykenæan pictures of "barnaculized" geese, and trees budding birds—two thousand years older than the Irish record, and nearly three thousand years earlier than the essay of the charming and persuasive professor.

CHAPTER XVI

SEA-SHELLS ON THE SEASHORE

ANY hard coat or covering enclosing a softer material is called a "shell"—thus we speak of an egg-shell, a nut-shell, a bomb-shell, and the shell of a lobster. But there is a special and restricted use of the word to indicate as "true" and "real" shells the beautiful coverings made for their protection by the soft, mobile animals called Molluscs. These animals expand and contract first this and then that region of the body by squeezing the blood within it (by means of the soft muscular coat of the sac-like body) into one part or another in turn. There is not enough blood to distend the whole animal, and accordingly one part is swollen out and protrudes from the shell, whilst another shrinks as the blood is propelled here or there by the compressing muscular coat. These creatures are the Molluscs, a name which has come into general use (and has even served as the title for a stage-play), as well as being the zoologist's title for the great division of animals which they constitute.

They are sometimes called "shell-fish," but this is no good as a distinctive name—since it is applied in the fish-trade to lobsters, crabs, and shrimps as well as to Molluscs. Lobsters, crabs, and shrimps are Crustacea, and totally different in their architecture and their mechanism from Molluscs. Familiar examples of Molluscs are the oyster, the mussel, the various "clams," and, again, the snails, periwinkles, whelks, and limpets. It is the shells of these animals which are "true" shells in the sense in which the word is used by "collectors" of shells, and in the sense in which we speak of "the shells of the seashore." These shells are usually very hard, solid things, made up of layers of lime-salts and horny matter mixed, and they remain for a long time undestroyed, washed about by the currents of the sea, and thrown up on to the beach, after the soft, oozy creature which formed them—chemically secreted them on its soft skin—has decomposed and disappeared. They are readily distinguished into two sorts—(1) those which are formed in pairs, or "bivalves," each member of the pair being called a "valve"; and (2) those which are single, or "univalves," often spirally twisted, as are those of snails and whelks, but sometimes cap-like or basin-like, as are the shells of the limpets. There is not so great a difference between bivalve and univalve shells as there seems to be at first sight. For if you examine the pair of shells of a mussel or a clam when they are quite fresh, you will find that the valves are joined together by a horny, elastic substance, and are, in fact, only one horny shell, or covering, which is made hard by lime deposited on the right and on the left, as two plates or valves, but is left soft and uncalcified along a line where these two valves meet, so as to allow them to move and gape, as it were, on an elastic hinge. It is the fact that the two valves of the shell of the bivalve, lying right and left on its body, correspond to the single shell of the snail or limpet, which differs from the bivalve-shell in not being divided along the back by a soft part into right and left pieces. That there is this real agreement between bivalve and univalve molluscs is quite evident when we examine the soft animal which forms the shell and is protected by it.

Though "shells" are often numerous on parts of the seashore, some beaches (as, for instance, at Falmouth, at the mouth of the Eden of St. Andrews, and at Herm in the Channel Islands) being so placed in regard to the currents and waves of the sea that great quantities of shells of dozens of species are thrown up, and even

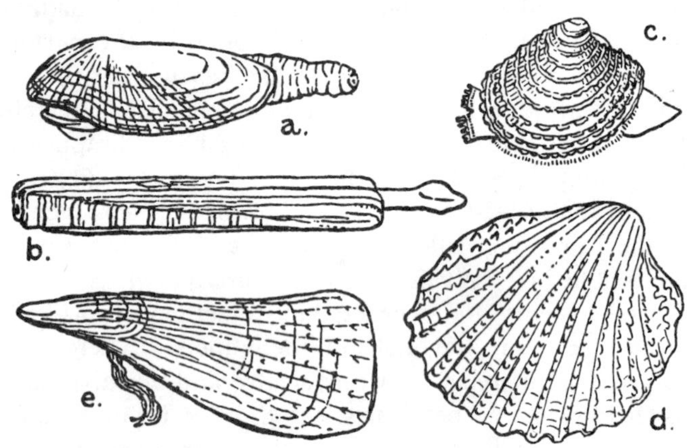

FIG. 19.—Some British Marine Bivalve Molluscs.

a, The smaller Piddock, Pholas parva, which bores into chalk, clay, and hard rock. Natural size.
b, The Razor-shell, Solen siliqua. The muscular foot is seen protruding from the shell. One-third the natural size, linear.
c, Venus verrucosa. Natural size.
d, Cardium echinatum. Two-thirds the natural size, linear.
e, Pinna pectinata, the "cappy longy." One-fifth of the natural size, linear.

"make up" the beach, yet there are not so very many Molluscs which live commonly on the shore between tide-marks. The shells which are accumulated as shell-beaches have come from animals which lived in quantity at depths of ten or twenty fathoms, whence they can be brought up alive by the dredge. There are, however, certain bivalves and certain univalves which are commonly to be found in the living state between tide-marks.

You will not find the oyster there on our own coast, but in Australia they have picnic parties where every guest provides himself with a hammer and a bottle of vinegar and a pepper-pot, and at low tide proceeds to chip the oysters off the rocks on which they grow tightly fixed, and to eat them "right away" before they have time to lose their good temper and sweetness! In Jamaica they show you oysters apparently growing on trees high up in the air, but they are dead, having attached themselves to the branches of a young tree which dipped into the water. Once fixed there, they were unable to move as the tree grew and carried them up with its branches above the sea-level.

The only bivalve at all common and visible to the eye between tide-marks is the common or edible sea-mussel, which is attached in purple clusters to the rocks (as in North Cornwall), or forms a wide-spreading pavement, called a "scalp," of as much as an acre in extent, on which thousands of mussels lie side by side. But by digging in the sand and mud between tides there are other living bivalves to be found, which burrow more or less deeply. The razor-shell (Fig. 19, b) is one of these (see p. 80). Often (as at Teignmouth and Barmouth) we find "cockles" buried in the sand, and those delicate, smooth bivalves not an inch long, white outside and purple within, which are made into soup at Naples and are called "vongoli," but have no English name. Other "clams" (Tapes, which is eaten in France, even in Paris, and Mya, and Scrobicularia which lives in black mud) may be dug up, but they are devoid of English names because we do not eat them; hence I have to speak of them by their Latin scientific names. As to univalves, there are three which are found almost everywhere on our coasts where there are rocks, namely, the periwinkles (one species of which actually lives above high tide-mark), the limpet and the dog-whelk. A small species of top-shell or trochus is also very common, and so is the chiton, or armadillo-shell, which, though really the most primitive and nearest representative of the ancestors of all univalve molluscs, yet has its own shell of a very peculiar character (sometimes with very minute eyes—true eyes—dotted about on it), and always divided transversely to its length (not right and left) into eight separate pieces, which, indeed, seem to be really separate, independent little shells, corresponding to eight segments like the segments of a shrimp or an earth-worm.

Let us now compare the soft animal of one of the bivalves—say the common cockle—with the soft animal to which a univalve shell belongs—say the limpet. They can be kept alive and watched in a finger-glass of sea-water, and can be removed from their shells and examined more closely—by killing them by dipping them for half a minute into very hot (not boiling) water. Both these molluscs—like all others—adhere tightly at one place to the shell. They cannot be removed from it alive, and make a new shell or creep back into the old one, as can some worms (*e.g.* the serpula) and other creatures which form a hard shell to live in. Certain muscles of the soft mollusc are so closely fixed to the shell that they must be torn in order to separate it. These muscles draw the two valves of the bivalve together, and shut it tight. You can verify this whenever the oyster-man "opens" an oyster for you. When at rest the shells gape, being kept open by the horny, elastic hinge-piece. Some bivalves (for instance, the common scallop, or pilgrim's shell, which can often be dredged in shallow water, and of which a large kind is sold in the London fish shops) actually swim in the sea-water by aid of this mechanism, the shells opening by elasticity and being closed by the muscle joining one to the other, at rapid intervals, flapping like the wings of a butterfly.

In the univalves the attachment of the muscle to the shell gives a fixed point for all the movements of the animal. The limpet has a well-marked head and neck—a pair of sensitive tentacles, and a small pair of dark-coloured eyes. The mouth is at the end of a sort of short snout. Just within the mouth, and capable of being pushed forwards to the level of the lips, is a most extraordinary rasp. It consists of a long ribbon, beset with fine horny teeth—very sharp and complicated in pattern. The ribbon extends far back into the body, and is worn away by constant use at the orifice of the mouth. It grows forward, like one of our finger-nails, as it wears out, and a new, unworn portion takes the place of that worn away. It is constantly in use to rasp and bring into the mouth the particles of the seaweed on which the limpet feeds. It is easy to remove this rasping ribbon with a needle or pen-knife, and examine it with a microscope. Every one of the hundreds of kinds of univalve molluscs has this ribbon-rasp, and its teeth are of different patterns in the various kinds. It is worked by very powerful little muscles, backwards and forwards, and is strong enough in the whelks to bore a round hole into other shells (for instance, that of the oyster), when the whelk proceeds to eat the soft animal, whose protecting shell has been thus penetrated. Some of the large marine snails produce a poisonous secretion from the mouth, which renders their attack with the ribbon-rasp all the more deadly to other marine creatures. The cuttle-fishes and octopods, which are molluscs too, possess, like the univalve limpets, snails, and whelks, this terrible ribbon-rasp in the mouth. It is an indication of a common parentage or ancestral relationship in the forms which possess it.

The cockle (Fig. 19, d), to which we now turn, has not got a ribbon-rasp, nor anything of the kind. It has a mouth with four flapper-like lips, but no projecting head, no eyes, no biting mechanism, nor have any of the bivalves, excepting a few which like the scallop have a series of eyes on the edge of the soft mantle or flap which lines the shell. This constitutes a greater difference between bivalves and the univalves than does the shape of the shell. They are a very quiescent, peaceful lot, feeding on microscopic floating plants (diatoms and such), which are drawn to the mouth by currents of water set going by millions of vibrating hairs arranged on four soft plates hanging under the protecting arch of the shell, and called in the oyster—in which bivalve most people know them—the "beard."

The limpet adheres to rocks by a great disk-like mass of muscle, which is called "the foot." It is really the whole ventral surface, and it can loosen its hold, and, by curious ripples of contraction, cause the animal to creep or glide over the rock. At low tide the limpet is exposed to the air, and remains motionless, but when the tide is up it makes a small excursion in search of food, never going more than a foot or two from the spot which it has chosen, and returning to it, so that in the course of time it actually wears away a sort of cup or depression at this spot—if the rock is not of exceptional hardness. The word "foot" is applied to the ventral disk-like surface of the limpet, because in many univalves this region becomes drawn out, and is connected by a comparatively narrow and nipped-in stalk or pillar with the rest of the animal. This occurs in the univalves which have large spiral shells, into which the whole of the soft animal can be deeply withdrawn, which is not the case with the limpet. You may find on the shore at Torquay a sea-snail (Natica), in which the animal is quite invisible, drawn far up into the shell. Place this in sea-water and watch it. Soft semi-transparent lobes begin to issue from the mouth of the shell, part of the soft distensible foot appears swelling out and growing bigger and bigger, and soft folds spread out from the mouth of the shell, and gently creep over it, and completely envelop it; the foot begins to grip the bottom of the vessel, and the animal "crawls." At last, swelling out from the other folds of soft but tense "molluscan" substance, the head and its tentacles emerge. Touch the animal and it shrinks rapidly, disappearing into the shell.

It used to be thought (about twenty-five years ago) that the molluscs expand their bodies in this manner by taking water, through definite apertures provided with valves, into their blood, and that, having thus swelled themselves out, they could shrink and reduce themselves by pouring out again the in-taken water. The behaviour of some other marine animals, namely the sea-anemones, which really do act in this way, made this explanation of the swelling and shrinking of molluscs seem probable. It was also known that the star-fishes and sea-urchins actually do take in the sea-water into a system of vessels connected with their wonderful sucker-bearing tentacles. But it turned out on close examination that the molluscs do not take in or shed out water in this way. A hole, which was thought to let in water into the blood of sea-snails, was shown to be only the opening of a great slime-gland. In the case of some bivalves which have red-blood corpuscles, I showed that the blood is never made paler, nor are the red corpuscles shed during the great distensions and contractions of the body. Measurements were made to determine the removal of water from a glass jar by an expanding sea-snail, and it was found that none is removed or taken up; in fact, the whole of what is very often an astonishingly large and bulky distension of the foot, or of lobes of the body, and the subsequent rapid shrinking of the same parts, depend entirely on the blood being injected from the rest of the body into the swelling part, and squeezed from it into the depleted region when the swollen part shrinks again. The firm, opaque shell hides from view the change of shape of the concealed body, and we see only the distended foot or other lobes which project from the shell.

The cockle has a "foot" of a very curious scythe-like shape, usually carried bent up between the two valves of the shell. Those who rightly like to confirm statements about unfamiliar animals can do so by buying a cockle or two at the fishmonger's. Some bivalves (the Noah's-ark-shell, called "Arca," and a few others) have a great flat foot, like that of the univalves, and crawl about on it. But in most bivalves it is curiously elongated and modified, for the purpose of burrowing into sand by vigorous strokes, and in some it is suppressed altogether, as in the oyster. The cockle is remarkable for the fact that when placed on a board or a rock it will give such a vigorous kick with its bent foot as to throw itself up a yard or so into the air. A naturalist (Stutchbury) dredging in Port Jackson, Australia, many years ago was overjoyed at discovering in his net three specimens of a very peculiar kind of cockle (Trigonia), which was till then only known in the fossil state from the oolite strata of Europe. He placed the three novelties on the seat of his boat, and was looking at other things when he heard a click-like sound, then another. He turned his head and saw that two of his newly-discovered "living fossils" had jumped overboard, and had the pleasure of seeing the third perform the same feat!

CHAPTER XVII

SAND-HOPPERS

WHEREVER there is a sandy seashore with here and there masses of dead seaweed and corallines thrown up by the waves, you will find sand-hoppers feeding on the debris. They are crustaceans, like crabs, shrimps, and barnacles, but in general aspect resemble enormous fleas. I hope that this comparison will not enable any reader at once to picture the less familiar by the more familiar. A good-sized sand-hopper is about half an inch long, and jumps not by means of a specially large pair of legs as the flea does, but by the stroke of the hind body, the jointed rings of which are carried curled downwards and ready to give a sudden blow. The sand-hopper (Fig. 20, a) has some of the rings or segments of the mid-body distinct, and not fused with those of the head or overhung by a great shield as in the lobster, crab, and shrimp. His walking legs and jaw-legs are also not quite of the same shape, though similar to those of a lobster, and his two little black eyes are not mounted on stalks, but are flush with the surface of the head. There are two quite distinct kinds of sand-hopper which live in crowds together on our sandy shores. They are not very different, but still are distinguished by naturalists from one another; one is called Talitrus (Fig. 20, a), the other Orchestia (Fig. 20, b). They are very similar in appearance and structure to a fresh-water creature common in weedy streams, which has no English name (except the general one of "fresh-water shrimp"), and is called by naturalists Gammarus.

In the open sea there are many hundreds of kinds of small crustaceans resembling the sand-hoppers in their compressed (not flattened) shape of body and in the details of their legs and the grouping of the joints of the body. Many of the smallest crustaceans which swarm in the surface waters of the sea and form part of that floating population, mostly of small transparent or iridescent and blue creatures, which we call the "plankton," or "surface - floating" population, and may be gathered by towing a very fine net behind a boat on a quiet day, can produce flashes of light which are vivid enough when seen at night. They contribute, together with jelly-fish and the teeming millions of minute bladder-like Noctiluca, and other unicellular animalcules, to produce that wonderful display seen from time to time on our coasts, and called "the phosphorescence of the sea." These minute crustaceans produce flashes of light by suddenly squeezing from pits or glands in the skin a secretion which is chemically acted on (probably

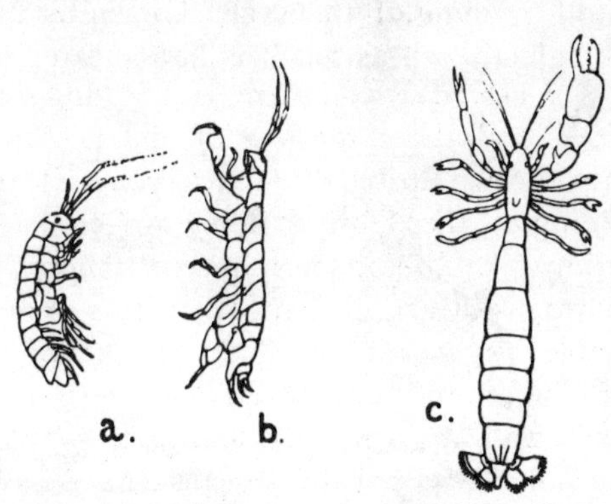

FIG. 20.—a, Talitrus locusta, b, Orchestia littorea, the two common kinds of "sand-hopper." Of the natural size. c, A kind of small lobster which burrows in the sand, Callianassa subterranea. About two-thirds the natural size, linear.

oxidized) by the sea-water, the chemical action setting up light-vibrations, but not the usual excess of heat-vibrations to which we are accustomed when light accompanies ordinary "burning" or "combustion."

Other crustaceans of several kinds, of an inch and more in length—transparent, delicate creatures, resembling small prawns in appearance—also produce light. Some of them are known by names referring to this fact, such as Lucifer (light-bearer) and Nyctiphanes

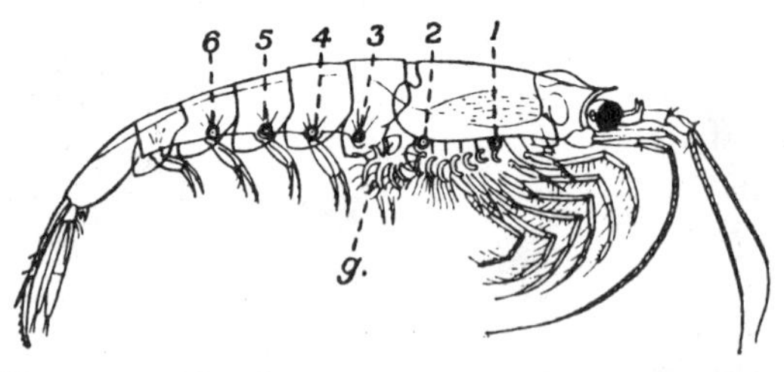

FIG. 21.—A Phosphorescent Shrimp (Euphausia pellucida). The lamp-like phosphorescent organs are numbered 1 to 6. There is another on the outer edge of the stalked eye, making seven in all on each side of the animal. *g*, points to the hindermost gill, enlarged.

(night-shiner). They possess special lantern-like knobs scattered about on the body, which have transparent lenses, and resemble small bull's-eye lanterns. Some have a row of seven lanterns on each side of the body (Fig. 21), but one kind has as many as 150 dotted about. These lanterns were only a few years ago thought to be eyes, and their elaborate microscopic structure was described as that of an eye. Of course, this was due to the fact that dead preserved specimens were studied, and not the living animal. Some twenty years ago I witnessed a most impressive exhibition of these phosphorescent shrimps at the house of my friend Sir John

Murray, of the "Challenger," at Millport, on the Clyde. He had obtained them (the kind called Nyctiphanes) in great quantities at a depth of ninety fathoms in the great Scotch fiord, and amongst other curious facts about them had shown that they enter Loch Fyne in vast numbers, and are the special nourishment of the celebrated Loch Fyne herrings. It had been noticed that the intestine of the plump, well-fed herrings is full of a deep-black substance, and Sir John Murray showed that this was the black, indigestible pigment of the eyes of the hundreds of phosphorescent shrimps swallowed by these favoured fish, which owe their fine quality to their special opportunity for feeding in the depths of the loch on the exceptionally abundant and nutritious light-producing crustaceans! At night my friend showed me a large glass vessel holding four or five gallons, in which were a hundred or so of the phosphorescent shrimps swimming around. We turned out the lamps of the room, and all was dark. Then a gentle tap was given to the jar, and each little crustacean lit up, as though by order, a row of seven minute lamps on each side of its body, swimming along meanwhile, and reminding one of a passenger steamer, as seen from the shore, as it glides along at night with its lights showing through a row of cabin windows. The shrimps' lights shone steadily for a minute or so, then ceased, and had to be lit up again by again signalling their owners by knocking on the glass. These little lamps, with their bull's-eye lenses, are far more elaborate structures than the glands which in other cases cause a flash by discharging a luminous secretion into the water. They are even more elaborate than the internal permanent phosphorescent structure of the glow-worm (an insect, not a crustacean), which has no condensing lens.

I have mentioned these phosphorescent organs of small and smallest crustaceans because not many years ago a French naturalist, my friend Professor Giard, found that many of the sand-hoppers on the great sandy shore near Boulogne are phosphorescent. A year or two later I found them myself on the shore above tide-mark at Ouistreham (Westerham), near Caen, where they had actually been mistaken for glow-worms! It was easy at night to pick up a dozen phosphorescent sand-hoppers during a stroll of five or ten minutes on the sands. Yet I have never seen them nor heard of their being seen on the English coast, and one of the results which I hope for in mentioning them here is that some of my readers will discover them on British sands and let me know. The remarkable fact about the luminous sand-hoppers is that they have no apparatus for producing light, and, as a matter of fact, do not produce it! Their luminosity is a disease, and is due (as was shown by that much-beloved teacher and discoverer the late Professor Giard) to the infection of their blood by a bacillus. Hence it is only here and there that you see the brilliant greenish ball of light on the sand due to a phosphorescent sand-hopper. And when you pick it up you find that the poor little thing is quite feeble and unable to hop. Examine its blood under the microscope and you find it teeming with excessively minute parasitic rods like those which cause the phosphorescence of dead fish, of stale bones, and occasionally of butcher's meat. Similar bacilli may be obtained by cultivation from any sea-water, and in such abundance that a room can be lit up by a bottleful of the cultivation. Perhaps all the light-producing bacteria or bacilli are only varieties of one species—perhaps they are distinct species. Whether a species or a variety, that which gets into the blood of the sand-hopper and gives it the luminosity of a glow-worm, inevitably and rapidly causes its death—a severe price to pay for brief nocturnal effulgence. Some of the germs can be removed on a needle's point from the dead sand-hopper and introduced by the most delicate puncture into a healthy sand-hopper or into a young crab, with the result that they too become illuminated, the bacillus multiplying within them. Being thus morbidly illuminated and having astonished the crustacean, not to say the human world, by their alarming brilliance, they quickly perish: a little history which may be read as a parable. The sand-hoppers give the disease to one another. It is, of course, a merely non-significant thing that the bacillus happens to set up light vibrations. Its chemical activity is concerned with its nourishment and growth, and in the course of these processes it not only produces light but poisonous by-products which kill its host. Some day we may get an "immune" race of sand-hoppers who will acquire the illuminating bacillus and defy its poison. Then we shall have a permanent and happy breed of brilliant sand-hoppers illuminating the dark places of the seashore.

It is conceivable that some of the disease-producing bacilli (bacteria, cocci, etc.) which multiply in man's blood and tissues should also produce light vibrations, and if one could be found that would render the blood luminous, whilst not producing much pain or *malaise*, no doubt some excuse would be found for its use as a fashionable toilet novelty. Cases are on record of luminosity of the surface of the body and its secretions being developed during serious illness by human beings, especially in acute phthisis; but these ancient records need confirmation.

Luminous bacilli or bacteria only give out light when free oxygen is in the water or liquid inhabited by them. A chemical combination of the oxygen with substances in the bacteria is the necessary condition of their evolution of light. When frozen, these bacteria cease to be luminous—the chemical combination cannot take place when the substance of the bacterium is frozen solid and maintained in that condition; the liquid condition is a necessary condition for these changes. These luminous bacteria have been used recently by Sir James Dewar in the Faraday Laboratory of the Royal Institution (where Sir James has shown them to me), for the purpose of investigating the action of intense cold on living matter. Although their luminous response to oxygen is arrested when they are frozen, yet immediately on allowing the temperature to rise above freezing-point the response of the living matter to oxidation recommences, and a luminous glow is seen. Hence we have in this glow a ready means of answering the question, "Does extreme cold, of long duration, destroy the simplest living matter?" Sir James Dewar has exposed a film of these bacteria to the extremest degree of cold as yet obtained in the laboratory, that at which hydrogen gas is solidified, and he has kept them in this, or nearly this, degree of cold for several months. Yet immediately on "thawing" the luminous glow was visible in the dark, showing that the bacteria were still alive. Curiously enough, whilst all chemical action in living matter can be thus arrested by extreme cold, and yet resumed on rise of temperature and restoration to the liquid condition, so that the old phrase and the conception of "suspended animation" are justified—yet there is one widely-distributed form of activity, the effect of which the bacteria, even when hard frozen, cannot resist, namely, that of the blue and ultra-blue rays of light. These rays, if allowed to fall on the hardest frozen bacillus, get at its chemical structure, shake it to pieces, destroy it. Hence Sir James Dewar argues that, whilst it would appear that the extreme cold of space would not kill a minute living germ, and prevent it passing from planet to planet, or from remotest space to our earth, yet one thing which is more abundant in space than within the shell of our atmosphere is absolutely destructive to such minute particles of living matter, even when hard-frozen, and that is intense light, the ultra-visible vibrations of smallest wave-length.

A dance on the seashore: a sketch by Edward Forbes (1852).

CHAPTER XVIII

A SWISS INTERLUDE

AFTER the hot summer of 1911 I escaped from London in September and made straight for Interlaken. Thence I was "wafted" by the electric railway to the "Schynige Platte"—a wonderful hill-side, 4500 feet above the "Bödeli," the flat meadowland in which Interlaken is placed. At the Schynige Platte we are separated to the south from the Jungfrau and the great Oberland range of mountains only by a deep rift in which rushes the "Black Lütschine," coming down from Grindelwald to join its "white" brother-torrent close beneath us at Zweilütschinen. To reach the "Platte" we creep in our train up the northern side of the mountain—one of whose peaks is known by the curious name "Gummihorn"—for more than an hour without a glimpse of what is on the other side. Then, when we are 6000 feet above sea-level, we enter a short tunnel in the shoulder of the mountain, and all is dark. When the train emerges every one in it gasps. You hear a cry from every mouth—for the scene is astounding! Coming through that tunnel we have stolen surreptitiously upon a band of gigantic snow-white brethren—the Wetterhörner, the Schreckhörner, the Eiger, the Mönch, the Jungfrau, the Mittaghorn, the Breithorn, and the Tschingelhorn. There they are—lying close to us, unaware of our approach—naked and unashamed, glistening in the sunlight, variously stretched in their immense repose. One feels on seeing them thus free from every scrap of cloud and clothing as though one had intruded upon a glorious company of titanic beings innocently sunning themselves in perfect nudity. It is with the sense that humble apologies for the intrusion are due to them, and will be graciously accepted because we hold them in such profound admiration and reverence, that we venture, little by little, to let our eyes dwell on their wondrous beauty. There are moments, it must be confessed, when we feel a qualm of modesty and are unwilling to take advantage of our rare chance—moments when we should not be surprised if one of the giants were to hurl a command at us—in terms of thunder and avalanche—ordering us at once to retire to the other side of the Gummihorn and leave them to their rightful privacy. There is no great view of snow mountains at close range—not even that from the Gornergrat—which is at once so fine and so easily accessible.

In the following year I went early in June in search of another Alpine delight, the spring flowers—not those of the highest "downs" and sheltering rocks 8000 or 9000 feet above sea-level, but those of the higher meadows, where the pine forests are beginning to thin out, and rich crops are cut before July by the skilful workers of the great Swiss industry, that of cow-herding and the production of cheese. It is difficult to define properly the term "Alpine" as applied to flowers. It is now used by horticulturists very generally for those exquisite small plants, the Saxifrages, Androsacæ, Gentians, etc., which grow in the highest regions to which plant-life extends—regions which are often covered by the winter's snow until June, and even late into that month. Some of these plants (as, for instance, the

Soldanellas—those little lilac-coloured flowers like pendent foolscaps which are allied to our primrose—and the crocus and the butterbur (Petasites) actually blossom beneath the snow and push their open flowers through it to the sunlight. Others of these "higher Alpines" have a peculiar mode of growth related to their special conditions of life. Their stems are very short and their foliage closely set, so that they form compact tufts or cushions, on which their short-stalked brilliant little flowers are dotted. The fact is they have not time in the short summer of these high regions to grow long stems. Their flowers are produced on low-lying parts of the plant, which carry small and abundant green leaves, but never send up long leaf-bearing stems. Not only do they thus do quickly, and without needless upward growth, what they have to do—namely, expose green leaves to the sunlight for nutrition and their flowers to the fertilizing visits of insects so as to ripen their reproductive seeds—but they benefit by keeping close to the warmth of the ground, which is heated by the strong sunshine, and is three and a half degrees higher in temperature than the cold moist air. In similar positions in low-lying regions the difference between the temperature of the air and that of the surface of the ground is not as much as one degree.

The Alpine meadows do not occur above the height of 5000 to 6000 feet, and are bordered by pine woods, in which are many beautiful plants not to be found at all or not in such profusion in the lower valleys. Both the meadows and woods of the Alpine heights graduate into those of lower level, and it is difficult to draw the line and say these flowers should be, and these should not be, called "Alpines." Many rock-loving plants allied to those found at great heights flourish in comparatively low-lying regions, where the necessary rocky character exists. The flowers of the high Alpine meadows are not the rock-lovers, the inhabitants of a surface formed by fragments of broken rock, to which the name "Alpine" is often limited. The meadow plants grow on good soil, and cover whole acres, in which there is but little grass. The fields are coloured of almost uniform blue or white or purple or yellow as the weeks go on, and various species one after another have their turn of dominance and maturity.

I paid, first of all, a brief visit to Aix and the lakes of Bourget and of Annecy, to the gorge of the River Fier, and to the finely-situated monastery of the Grande Chartreuse—a huge building, devoid of beauty, which it seems to be difficult to utilize now that the Carthusian Brothers have been expelled. The richly-coloured Alpine centaury, deep blue and purple red, was growing in the woods around it abundantly, and many other handsome plants. Zoology was represented by most excellent little trout provided for us at the village inn. Then I stayed a couple of days at Geneva, where, in a pool in a richly-planted rock garden—that of the well-known horticulturist M. Correvon—I came across what I have long wished to see, namely, the blue variety of the edible frog. Six years ago I wrote an account of the little blue frog of Mentone, the rare variety of the green tree-frog, or rainette, so abundant in that region (see "Science from an Easy Chair," p. 50: Methuen, 1910). The edible frog (Rana esculenta) is often very beautifully coloured with blotches of dark brown and pale green, and a pale yellow stripe down the back. It is easily distinguished from the brown frog (Rana temporaria), which occurs with it. The latter is the common frog of our islands, though we also find the edible frog in the South of England. The blue variety of the edible frog has been seen in various localities in Germany and along the valley of the Rhone. It owes its colour, as does the blue tree-frog, to the suppression of yellow pigment in its skin. The one I found was swimming in a small clear pool with two other very finely-marked specimens of the more usual colouring. A blue variety of our common brown frog has not been observed, although it is occasionally very pale in colour and, on the other hand, is sometimes of a bright orange-brown tint. Several species of toads and frogs are found on the Continent which do not occur in Great Britain.

Years ago (when France and Germany began the great war of 1869–70) I travelled from Geneva to Chamonix by coach. It took the whole day. Now I and my companion, avoiding the railway, were driven in a motor-car past Bonneville, Cluses, and Sallanches (with its famous view of Mont Blanc), and along the vale of Chamonix to its far end above Argentière in less than three hours. Here we stayed a few days in the Hôtel du Planet, at a height of 4500 feet, in order to enjoy the sight of the meadows and woodland flowers. I may add that in this quiet hotel the proprietor gave us simple, good food, well cooked, which is more than I can say of the large hotels on the lakes and popular resorts, such as Geneva, Montreux, Glion, and Interlaken, where I have carefully inquired into the kitchen arrangements and food supplies. The latter barrack-like edifices have of late years become intolerable owing to the mechanical supply to them (by a group of monopolist financiers who have acquired the contract) of the nastiest ice-stored fish, meat, and vegetables. These are heated in their kitchens with bottled sauces in patent ovens by underpaid scullery-helps, without the superintendence of a qualified "cook." The result is a sham—pretentious and inedible—which yields a fine profit to the hotel companies, and is erroneously believed by the travelling crowds of to-day to be French cookery! In reality it is a new device for bringing the "catering" in all hotels in the great holiday centres under a monopolist control. The scheme is similar to that to which the continental railway companies have yielded in leasing to a well-known company the restaurant and sleeping arrangements on their trains, with the result of causing much misery to travellers and profit to themselves and to the monopolists.

Owing to differences in exposure and soil, the meadowland above Argentière showed a fascinating variety of colour. Here was an acre of the large-flowered purple geranium, interspersed with the big Alpine yellow rattle (a greedy root-parasite); there (near some pine trees) a mass of the yellow anemone (Anemone sulfurea); farther on a whole meadow, blue with the abundance of large hairbells and viper's bugloss. Close by, in the damper parts of the valley descending from the Col des Montets, three or four acres of meadowland were white, so thickly were they covered with tall plants of the distinguished-looking white buttercup (Ranunculus aconitifolius). In some parts, among these dignified Ranunculi, the plump yellow heads of the globe-flower (Trollius), also a kind of buttercup, were abundant. Overshadowed by these larger plants, or growing up between them, were orchids, plantains, polygonums, and many others. The most beautiful plant in these meadows was St. Bruno's lily, which we found in abundance on a steep bank. It is named after the founder of the Carthusian order, whose monastery (the Grande Chartreuse), first established when William the

Conqueror ruled England, I had visited a week before. St. Bruno's lily has large, white, funnel-shaped flowers, an inch or more long, three or four on a stalk. It is known to botanists by the pretty name "Paradisia liliastrum." It is the lily of the Alps, pure and un spotted, with a delicious perfume, and six golden stamens guarded by its beautiful and large white corolla. In the woods we found some of the larger orchids, and also whole banks covered with the waxy-looking flowers, variegated in colour, white, yellow, and red, of the large millwort, the Polygala chamæbuxus—a plant very unlike in appearance to the little blue and white milk-worts of England. It flowers in winter as well as through the early summer. Another wonderfully waxy-looking flower which we found is that of the shrub known as the Alpine Daphne. There is something suggestive of exotic rarity and perfume about a waxy-looking flower. Of the same character are the flowers of the little shrubs of the genus Vaccinium known as the bilberry, the wortleberry, the cow-berry, and the bear-berry, which occur on the open scrubland. The rusty-leaved Rhododendron, with its crimson flowers, and the little Azalea (like the Vaccinia—all members of the Heath family) were abundant—as well as the true dark-red rose of the Alps, the richly-scented Rosa alpina.

We left Argentière and the constant companionship of the great glaciers of the vale of Chamonix, and descended by train through the awe-inspiring valley of the Trient (up which we used to walk many years ago, on our way to the higher regions) to Martigny, and then drove for four hours up a rough mountain road to the hotel of Pierre-à-voir—whence we descended a few days later in sledges, over grass slopes and torrent beds,

4000 feet in an hour and a quarter, to Saxon in the Rhone valley, a truly alarming experience. The "luge" or sledge is supported in front by a strong mountaineer who prevents it from "hurtling" down at breakneck speed, topsy-turvy. As the avoidance of such a catastrophe depends on the strength and the sureness of foot of this individual, travelling by "luges" is not to be recommended in summer, however agreeable it may be when the mountain side is covered with snow. In the woods near Pierre-à-voir we found another member of the Heath family, looking like a lily rather than a heath, the sweet-scented winter-green with its large single white flower (Pirola uniflora), and on the rocks on open ground masses of the pink flowers of the little rock soap-wort (Saponaria ocymoides). The curious tall, big-leaved composite with only three purple florets to a head, the Adenostyles albifrons, was here much in evidence. We were too early for the flowers of the pretty little creeping plant allied to the honeysuckle which the great Linnæus asked his friend Gronovius to name after him, the Linnæa borealis, though we had been told that it grows in this neighbourhood.

Then we spent five days at Glion and on the incomparable Lake of Geneva, never wearied of gazing at the changing mysterious lights and colours (sapphire, emerald, and silver) of its vast and restful expanse.

The question often is asked, "Why is it that the same species of flower is brighter and stronger in colour when growing high up in the Alps than when growing in the lowlands and in our own country?" The fact is admitted; the blues of the blue-bells (Campanula), the bugloss, the forget-me-nots, the crimsons and purples of the geraniums and the pinks and the campions, and many others, are examples. Careful study and consideration of the facts have enabled botanists to show, in many instances, within recent years, that the peculiarities of form and also of colour of the stems, leaves, and flowers of plants are not mere unmeaning "accidents," but are definitely of advantage and of "survival value" to the species. Thus we have seen that the tuft-like cushions formed by high Alpine plants are explained. The purple and reddish colour of stalks and leaves like that of the red variety of the common beech has not always, as in that plant, the purpose of protecting the chlorophyll from destruction by too vivid sunlight. In Alpine plants it is often present on the underside of leaves and of the petals, and acts to the plant's benefit, absorbing light and converting it into heat. But it also seems in many cases to protect the juices of the plant from the destructive action of white light.

It is held by some botanists that the bright colour of Alpine (and Norwegian) samples of a flower elsewhere of a paler colour is due to the direct action of the greater sunlight of the high regions in causing the formation of pigment. This is inadmissible. The sunlight cannot act in that way. It causes increased formation of nutriment by acting on the chlorophyll, and an Alpine plant thus highly charged with nutritive matters can afford to form more abundant pigment than a plant which enjoys less brilliant sunshine. The high-coloured Alpine flowers are a breed or race; a pale-coloured plant taken to the Alps from below does not itself become high coloured. It is a matter of natural selection. The occasional high-coloured "spontaneous" variations produced from seed have an advantage in the short summer of the high Alps. They attract the visits of the few insects in the short season more surely than do the paler individuals, and consequently they are fertilized and reproduce, whilst the race of the paler individuals dies out from failure to attract the insects. Thus we get a high-coloured race established in the mountains, a race that can make haste and seize the brief opportunities of the short but brilliant summer. There are many peculiarities of form and colour of plants the life conditions of which are diverse (e.g., woodland, moorland, aquatic, seashore, dry air, moist air, etc.), which can be shown by accurate observation to be specially related to those life conditions. Those conditions allow the peculiarities to survive and establish a race, in some cases a species, whilst preventing the maturity or destroying the life of those individuals not presenting that advantageous peculiarity of variation.

CHAPTER XIX

SCIENCE AND DANCING

THERE is at the present day in this country a real and most happy revival of interest in the great art of dancing as exhibited on the stage. We owe this to the creative ability of the musical composers and directors of the Russian Imperial Ballet, as well as to the highly-trained and gifted Russian artists who have visited this country, and especially to the poetical genius of Madame Anna Pavlova. Though dancing may seem, on first thought, a subject remote from science, yet, like all other human developments, it is a matter for scientific investigation, and one upon which science can throw much light. What is the origin and essential nature of "dancing"? Do animals dance? What is its early history in mankind? What is its relation not merely historically, but from the point of view of psychology—the study of the mind—to other arts? What is its real "value" and possible achievement?

To dance is to trip with measured steps, and, whilst primarily referring to human movement, the word is secondarily applied to rapid rhythmic movements even of inanimate objects. Rhythm is what distinguishes dancing from ordinary movement of progression or from simple gesture or mere antics. Dancing on the part of man or animal implies a sense of rhythm. Though not common amongst animals, it is exhibited by many birds, by spiders, and by some crustaceans! Rhythm is an essential feature of the sequence of sounds which we call "music." The singing of birds is related to their perception of and pleasure in rhythm, and it is not, therefore, surprising that they should also dance. It is, however, curious that the birds which "dance" are not the "singing birds," and that there are many birds which neither sing nor dance. The dancing of birds is usually part of the "display" of the males for the purpose of attracting the females at the breeding season. It is well known in some African cranes, as well as in rails and other similar birds, and may be witnessed at the Zoological Gardens in London. Other birds "strut" rather than dance, whilst displaying their plumage, as, for instance, the turkey and pheasant tribe and the bustards. Parrots and cockatoos will often make a rhythmical up-and-down movement of the neck in time to music, but usually the "dance" is the accompaniment of definite emotion. The male spider of some species courts the female by making dancing movements and posing itself in a very curious way, so as to display a spot of bright colour on the head to her observation. The same kind of movement and action has been observed in marine shrimp-like creatures. Some spiders are excited and made to dance by the vibrating note of a tuning-fork set going near them. I once had the chance to observe a male octopus in the aquarium at Naples, who was displaying himself to the female, changing colour rapidly from one shade to another, and rolling his long sucker-bearing arms in the form of spirals. Probably one should not consider this as a "dance," since no rhythmic interruption or succession of movements was observable.

It is established that in mankind, as well as many animals, when in a state of emotion, movement and gesture, as well as the vocal utterance, take on a rhythmic character, that is to say, become a dance and a song. The emotion is not necessarily that of amorous passion; in mankind it is frequently of a warlike or religious character, and is worked up by the sympathy, imitativeness, and desire for unison in expression which is common in troops or large gatherings of animals of social habits. Man presents a more advanced development in variety, sensitiveness, and abandonment to social or combined action and expression than do other animals, and this is equally true of the more civilized and of the more barbarous races. Apparently in obedience to the same tendencies as those which convert simple forms of movement into a rhythmic dance, the speech of man, under conditions of emotion, assumes a rhythmic form, so that dancing bears the same relation to the ordinary movements of locomotion and gesture which verse does to ordinary speech, or, again, which song bears to mere exclamations and cries, indicative of feeling. Dancing is the universal and most primitive expression of that sense of rhythm which is a widely distributed attribute of the nervous system in animals generally. In primitive men it is a simple but often very violent demonstration of strong emotion, such as social joy, religious exaltation, martial ardour, or amatory passion. The voice and the facial muscles, as well as those of the limbs and body, are affected, and the dancers derive an intense pleasure from the excitement, which so far from exhausting them leads them on to more and more violent rhythmic or undulatory action. In its purest form this ecstatic condition is seen in the spinning dervishes. It was developed into the mad and dangerous festivals of the worshippers of Bacchus and other deities in ancient

Greece. It has been seen in mediaeval Europe as the dancing mania and tarantism. The liability to this and similar forms of "mania" lurks beneath the surface among populations which are nevertheless staid and phlegmatic in their usual behaviour. The Romans in ancient times recognized its unhealthy character, and though fond of ceremonial dances and theatrical shows, and even of the performances of dancing girls from Greece and the East, disapproved of dancing on the part of a Roman citizen. Cicero says, "As a rule no one, who is not drunk, dances—unless he is, temporarily, out of his mind."

Although the mad performances of bacchanalians and dervishes are recognized as unhealthy, civilized peoples in Europe since the fifteenth century have developed and practised dancing as an art in two directions—first, as a popular amusement in which definite combinations of graceful movements are performed for the sake of the pleasure which the exercise affords to the dancer and to the spectator, and secondly, as carefully trained movements which are meant by the dancer vividly to represent the actions and passions of other people, and are exhibited by specially skilled performers on a stage. The first kind is what we call "country dances," "popular dances," also "Court and ball-room dances," and has been commended by the philosopher Locke and other writers as a valuable training for both mind and body, and by physicians as a health-giving exercise. The second is "the ballet."

In the dances of savages and primitive peoples, some kind of music is always found associated with dancing, the one helping and developing the other; they are descendants of one parentage. Very commonly, too, some kind of "acting"—the representation of a hunt, a fight, or a love adventure—is an important feature of such dancing. Modern popular and Court dances are intimately connected with and dependent on special music, the rhythm and variation of time and strength in which is, as it were, illustrated by the dancing, and serves to guide it and to keep the dancers in unison. The signification behind all such modern dancing is courtship—the addresses of the man to the woman, and her elusive reception or rejection of them. In the Cathedral of Seville, however, you may still see, at the festival of the Corpus Christi, a religious dance, a dance of worship and adoration, performed by acolytes in front of the high altar. In the early days of the Church such ritual dancing, by both old and young, was a regular thing, as it was in the still earlier religious ceremonies of the ancient Romans and in the time of King David.

The development of dancing as a fine art has only been rendered possible by the establishment, under the patronage of various European princes, of great exhibitions of dancing, called "ballets," and the creation of a profession of dancers, who, like professional actors and musicians, devote their lives to the study of their art and the training necessary for efficiency in its practice. In this, its highest development, dancing, whilst maintaining its dominance, is entirely dependent on the aid of music, and becomes blended with the art of the actor and pantomimist. As in "opera" the effect of the musical art is enhanced by the meaning of the words sung, by the acting of the performers, and by the accessories of scenery and costume, so in the ballet do all these factors, except the human voice, contribute to the artistic result. The latest development of the ballet is, in fact, "grand opera," without a voice, without words. Gesture, facial expression, and movement of the limbs, marvellous for its grace and directness of appeal, take the place of words. In fact, dance, the appeal to the eye, takes the place of verse, the appeal to the ear. And it is a fact, unexpected and astonishing to those new to it, that the same quality of "poetic imagination" which distinguishes "word-poems" from mere doggerel or commonplace verse, can also inspire the great dancer and give to a wordless dance the unmistakable value of poetical art, distinguishing it from purely acrobatic or barbaric capering. It is a fact that poetic imagination may be conveyed in one kind of art as in another, and that dancing, though greatly limited in its range of detailed expression, yet is closely similar in its forms to music, verse, and to glyptic and pictorial art, of all of which it is the parent and forerunner. Its primitive character is no less remarkable than the readiness with which it exerts its charm and develops new importance at the present day.

Regarded as a fine art, and not merely as a pastime, dancing has frequently great beauty in its simple quality of the rhythmic movement of decorative form and colour. The dances depicted on Greek vases had this character, and so, with varying degree of merit, have the ballets common during the last fifty years in London and other great centres. But before this period the makers of ballets (a word originally signifying to dance, to sing, to rejoice, and representing three modern words—ballet, ball, and ballad) did not aim at a mere exhibition of living rhythmic decoration, but at the production of a theatrical performance in which a story is told only by gesture and dancing accompanied by music. The real modern founder and exponent of the ballet as thus understood was Noverre, a Frenchman (called by Garrick "the Shakespeare of the dance"), who died in 1810. He brought to a high degree of perfection the art of presenting a story by pantomime, and he never allowed dancing which was not the direct expression of a particular attitude of mind. His professed effort was to introduce the steps and poses of ancient Greek dancing shown in sculpture and painted pottery—as *the* model for stage dancing. And he succeeded. The great dancers of the past who are known to us by tradition—Vestris, Camargo in the eighteenth, and Cerito, Grisi, and Taglioni in the earlier half of the nineteenth century—were not merely perfectly trained as dancers, but were actors, and possessed poetic imagination. Women did not appear in the ballet until the time of Louis XIV, and Mlle Camargo was the first to wear the conventional short stiff ballet skirt.

"Convention" has a great weight in such matters. But it seems to be undeniable that the conventional ballet-skirt conceals the beautiful movement of the leg on the hip joint, a disadvantage from which the male dancer does not suffer. Skirts are, in fact, out of place in really fine dancing. Flowing light drapery, or better still the Circassian jacket and full gauzy trousers fastened at the ankles, are the only possible dress for a really great *danseuse*.

The dramatic ballet or *ballet d'action* lasted until the end of the fifties in London, and then ceased almost suddenly to occupy the leading position which it once held at the Opera House. In London, as in Paris and Vienna, it was transformed into a mere spectacular display of costume and meaningless rhythmic drill. The dramatic ballet ceased to exist. The great tradition of fine stage-dancing and ballet-drama was, however, preserved in Russia. It is not easy to explain, but the fact is that two peoples so far apart as the Russians and the Spaniards are more devoted to dancing than any other European nationalities. Successive Tsars have spent large sums in maintaining colleges in St. Petersburg and in Moscow, where boys and girls are lodged and carefully educated whilst they are trained from the age of ten years in the art of stage-dancing. The greatest musical composers have been encouraged to write "ballets," and the ablest designers and "producers" have been secured by large salaries. Something like £80,000 a year is spent by the Tsar on the maintenance and development of this beautiful art, which is dead elsewhere, but seems to fit the genius of the Russian people. A new respect for Russia, a profound admiration for the Russian artists, has been the result of the revelation of the Russian ballet by the recent visits of its members to this country.

During the last thirty years of its period of nurture and development in Russia the ballet has developed in two directions. Neither of these are popular and successful in Russia, where the old traditional and established ballet of the early nineteenth century—what may be called "academic" dancing—is alone in demand. What *we* call "the Russian ballet" is dramatic in nature, and includes such wonderful combinations of music, scenery, costume, and perfect artistic expression by dancing and gesture as we have seen in Scheherazade, Cleopatra, Prince Igorre, Tamar, and Petrouschka. It promises in its latest development to supplant the musical drama known as "opera," in which the human voice is used. But the most striking development is that in which dancing appears as the exponent of lyrical poetry. It is to the teaching of Isadora Duncan that the Russian dancers admit their indebtedness for this new departure. When undertaken by untrained dancers and amateurs (even by the innovator herself) the attempt to interpret lyrical subjects showed some ingenuity in conception, but failed to command general appreciation, as the efforts of a painter or an actor, who has not acquired command of the material of his art, also fail. But when Anna Pavlova brought her lifelong training as a dancer and her poetic imagination to the interpretation of master-pieces of music inspired by such subjects as "Night," "The Dying Rose," "The Wounded Swan," and the moonlight mystery of "Les Sylphides," a new and most poignant form of emotional expression became apparent. A single figure moving over the stage with expressive steps and gestures of the arms, with lips and eyes guided and controlled by consummate art, blended itself with and interpreted to the spectator the poetic thought of a great musical composer and a great writer. This new development of the dancer's art may remain with us. But it requires the presence of one who combines the rare gifts possessed by Madame Pavlova—perfect technique and poetic sympathy.

Many people derive a definite part of the pleasure given to them by an orchestral concert from the contemplation of the movements of the instrumentalists and the directive interpreting gestures of a great "conductor." Others would prefer the orchestra and its leader to be unseen; they find special delight in hearing great music surge and float from no visible source through the dimly-lit aisles of a vast cathedral. They do not desire their eyes to be called in aid of music unless the appeal to vision is complete and worthy of the theme. It is, I think, undeniable that Dr. Richter and my friend Sir Henry Wood, whose expressive backs and persuasive hands are so dear to concert audiences, are a kind of dwindled ballet dancers, connected by the drum-major of the military band and the dancing "choragus" with the primeval phase of the arts when music and dancing were inseparable.

CHAPTER XX

COURTSHIP

IT is always amusing to find the lower animals behaving in various circumstances of life very much as we do ourselves. There is a tendency to look upon such conduct on the animals' part as a more or less clever mimicry of humanity—a sort of burlesque of our own behaviour. Really, however, it has a far greater interest; it is a revelation to us of the nature and origin in our animal ancestry of various deeply-rooted "behaviours" which are common to us and animals. The wooing of a maid by a man and the various strange antics and poses to which love-sick men and women are addicted, are represented by similar behaviour among animals, and that, too, not only among higher animals allied to man, but even among minute and obscure insects and molluscs. In fact, the elementary principle of "courtship" or "wooing," namely, the pursuit of the female by the male, is observed among the lowest unicellular organisms—the Protozoa and the Protophyta—and it holds among plants as well as among animals, for it is the pollen—the male fertilizing material—which travels, carried by wind or by the nectar-bribed "parcels-delivery company" of bees, to the ovules of a distant flower, and not the ovules (the female products) which desert their homes in quest of pollen.

The "reproduction," or producing of new individuals, of many animals and plants can be, and is, effected by the detachment of large pieces of a parent organism. Thus plants split into two or more pieces, each of which carries on life as a new individual. Many worms and polyps multiply by breaking into two or more pieces, and very often the broken-off pieces which thus become new individuals and carry on the race are extremely small, even microscopic in size. The spores of ferns and the minute separable buds of many plants and animals are of this nature. They grow into new individuals without any fusion with fertilizing particles from another individual. Yet there seems to be even in the very simplest living things a need to be met, an advantage to be gained, in the fusion of the substance of two distinct parents in order to carry on the race with the best chance of success. We find that those organisms which can multiply by buds and fission yet also multiply regularly by ovules fertilized by sperms. We see this process in its simplest condition in microscopic plants and animals which are so minute that they consist of only a single "cell"—a single nucleated particle of protoplasm. Such unicellular organisms have definite shape, even limb-like locomotor organs, shells, contractile heart-like cavities within the protoplasm, even mouths, digestive tract, and a vent. They produce new individuals by merely dividing into two equal halves or by more rapidly dividing into several individuals each like the parent, only smaller. But from time to time, at recurring periods or seasons, two of these unicellular individuals (of course, two of the same kind or species) come into contact with one another, not by mere chance, but attracted and impelled (probably by chemical guiding or alluring substances of the nature of perfumes) towards one another, and then fuse into one. Two (or sometimes several) individuals thus melt together and become one individual—a process the exact reverse of the division of one into two. This is known to microscopists as "conjugation." The new individual resulting from conjugation after a time divides, and the individuals thus produced, each consisting of a mixture of the fused and thoroughly mixed substance of the two conjugated individuals, feed and grow and divide in their turn, and so on for several generations, until again the epidemic of conjugation sets in, and the scattered offspring of many distinct pairs of the previous conjugation-season in their turn conjugate.

It is clear that the tendency of this process is to prevent the continued multiplication of one stock or line of descent in a pure state. By conjugation different lines of descent—the progeny of different individuals, often brought together from widely separate localities—are blended and fused. And this is, we are led to conclude, a matter of immense importance. To effect this mixture of separate stocks is, as Darwin has shown, a prime purpose of the habits and structures implanted in the very substance of living things, and developed and accentuated in endless ways and with extraordinary elaboration of mechanisms and procedure during the immense lapse of ages during which life has unfolded and developed on this earth. The fusion of different strains by conjugation gives increased variation in the offspring or new generations: for the two parental strains differ more or less, as all living individuals do, from one another. The result of their fusion is different from either parent. In fact, the process of fusion itself causes a disturbance—a readjustment of the living matter—so that completely new variations result and are selected or rejected in the struggle for existence. Either parental strain was perhaps not so suitable to a newly developed change in the surrounding conditions of life as the new blend may be. Thus a more certain and active production of possibly useful variations is provided for than would be the case were the variations of one self-multiplying stock alone presented for selection.

In the case of simple conjugation the cell individuals which fuse or "mate" with one another, and may be called "maters" or "mating cells," are in all respects similar to one another. But we find among the unicellular plants and animals cases in which one of the mating cells, instead of fusing with another straight away, divides into a number of much smaller cells, which are very active in locomotion and are specially produced in order to mate or fuse with the larger cells. The mating cells are called "gametes," and the large motionless mating cells are called "macro-gametes," or "large maters," whilst the small motile mating cells are called "micro-gametes," or small "maters." The former are of the same nature as egg cells or ovules, the female reproductive particles, whilst the latter, the small "maters," are identical in nature with the sperms or spermatozoa or male reproductive particles of higher organisms. In the case of certain parasitic unicellular animals called coccidia, and also in the parasite which causes malarial fever, quantities of small "mating cells" are produced which fuse with or "fertilize" other much larger mating cells. The small "maters" of coccidia have long vibrating tails and minute oblong bodies, and agree closely in appearance and active locomotion with the spermatozoa of higher animals and plants. The large spherical mating cells might be mistaken for the egg cells of larger animals. In the globe animalcule, Volvox globator, we find a transitional condition leading us to the production of small (male) and large (female) mating cells, like those regularly produced by the massive plants and animals which are built up by hundreds of thousands of "cells" or protoplasmic units conjoined and performing different services for the common life. Volvox is one of those simple aquatic organisms which is not a single cell but a group of many cells (some hundred) hanging together—in this case so as to form a hollow sphere. All the cells of an individual sphere are alike, and have originated by division from one first cell. When the "breeding season" arrives one or two cells of the sphere increase in bulk—they become "large mating cells"—in fact, egg cells. At the same time one or two divide (without separating), so as to form packets of minute oblong cells with vibrating tails. These are "small maters," or "spermatozoa." When ripe they separate and swim away to fertilize—that is to say, to fuse with—the large "mating cells" or egg cells of other Volvox spheres. Such a Volvox sphere as I have described is "bi-sexual": it produces both large and small mating cells, both male and female reproductive cells. But sometimes we find that a number of Volvox spheres produce only large mating cells by the swelling up of one or two of their constituent cells. They are, in fact, female Volvox spheres. And other Volvox spheres produce only packets of small mating cells by the splitting and change of one or two of their constituent cells. They are male Volvox spheres.

When we now look at the higher plants and animals formed of aggregations of innumerable cells (all derived from the division of a first cell—an embyro cell or fertilized egg cell) we find that amongst the mass of variously shaped cells forming the "tissues" of these higher organisms some are set apart even in early growth as "mating cells" (gametes or reproductive cells).

Usually they are in two groups—namely, the ovary, which includes the large mating cells or egg cells or ova; and the spermary, which includes the cells which break up into small mating cells or sperms. In many animals both ovary and spermary are present in the same individual, but in most of the larger animals (insects, crustaceans, and vertebrates) either the ovary is suppressed, when the creature is called a male, and produces only small mating cells, or the spermary is suppressed, and the creature is a female, producing only egg cells. In both cases there may be a distinct but minute representative of the suppressed organ present and recognizable by its microscopic structure.

The point in this history, which seems to be important and must not be lost sight of, is that the small mating cell is in all the stages cited actively mobile and swims rapidly through water when its producer is an aquatic animal. The large mating cell is quiescent. It is more or less swollen with granular nutrient particles—often vastly so enlarged. It already is acting the maternal part, preparing nourishment for the growing embryo which will develop from its protoplasm when fused with that of the relatively tiny but active male mating cell. And it is certainly very noteworthy that when these two kinds of mating cells become separated in distinct "carriers" (that is to say, produced one without the other in what are called male and female individuals), the primitive character of the mating cells—whichever of the two kinds they be—impresses itself on the complex elaborate many-celled organism in which they arise. The male is the more active, the more disposed to travel. It is always the male who seeks, courts, woos, and attacks the female, as the small mating cells seek and attack the larger mating cells. The character and conduct of the female animal is largely (not without deviations and additions) based on that of the larger mating cell or macro-gamete; she is the one who waits, is sought, is courted, and wooed. And like the egg cells of which she is the vehicle and envelope, she is specially concerned in the provision of nutriment for the early growth of the young.

Courtship, then, seems to have had its foundations very deeply laid, even in the earliest and simplest forms of life—at the time when the principle of the union of the substance of two strains to produce a new generation was established, and when, further, the active, seeking male cell was differentiated from the immobile nourishing female cell.

Amongst the polyps, sea-anemones, and jelly-fish, though we frequently find that there are distinct males and females, there is no courtship. This is connected with the fact that, like plants, they are (excepting the jelly-fish) fixed and immobile. The male cannot "court" the female, because neither of them can approach the other. I once saw in the aquarium at Naples a sudden and simultaneous discharge of a white cloud, like dust, into the water from half the magnificent sea-anemones fixed and immobile in three large tanks. The cloud consisted of millions of the small "mating cells," and were thrown off by the males. They were carried far and wide by the stream running through the tanks. In the sea such a discharge would be carried along by currents, and might fertilize egg-bearing sea-anemones of the same species growing a mile or two away.

It is when we have to do with actively moving animals that "courtship" comes into existence. It has many features and phases, which comprise simple discovery of the female and presentation of himself by the courting male; attempts to secure the female's attention, and to fascinate and more or less hypnotize her, by display of brilliant colours or unusual and astonishing poses or movements (such as dancing) on the part of the male; efforts of the male to attach the female to himself, and deadly, often fatal, combats with other males, in order to drive them off and secure a recognized and respected solitude for himself and his mate. The courtship of many insects, crustaceans, molluscs, fishes, reptiles, birds, and mammals has been watched and recorded in regard to these details. Naturally enough, it is in the higher forms, the birds and the mammals, that there are the most elaborate and intelligible proceedings in regard to the attraction of the female. But when we compare what birds or, in fact, any animal, does with what man does, we must remember that man has, as compared with them, an immense memory, and has also consciousness. All other animals are to a very large extent mere automata, pleasurably conscious, perhaps (in the higher forms), of the passing moment and of the actions which they are instinctively performing, but without any understanding or thought on the subject. They cannot think because, though some of them are endowed to a limited extent with memory, they have not arrived at the human stage of mental development when consciousness takes account of memory, a memory of enormously increased variety and duration.

Man has more and more, as he has advanced in mental growth, rejected the unreasoning instinctive classes of action, and substituted for them action based on his own experience and conscious memory, action which is the result of education—not the education of the school, but that of life in all its variety. But in many things he is still entirely guided by unreasoning mechanical instinct, and in others he is partly impelled by the old inherited instinct, partly restrained and guided by reason based on experience and memory. This makes the comparison of the courting man with the courting animal doubly interesting. We ought to distinguish what he is doing as a result of ancient inherited mechanism from what he is doing as a result of conscious observation, memory, and reasoning.

CHAPTER XXI

COURTSHIP IN ANIMALS AND MAN

THE German poet Schiller arrived long ago at the conclusion that the machinery of the world is driven by hunger and by love. If we join with hunger, which is the craving of the individual for nourishment, the activities which aim at self-defence,—whether against competitors for food, against would-be devourers, or against dangers to life and limb, from storm, flood, and temperature,—we may accept Schiller's statement as equivalent to this, namely, that the activities and the mechanisms of living things are related to two great ends—the preservation of the individual and the preservation of the race. "Love," or what we should call in more discriminating language "amorousness," or the "mating hunger," is the absolute and inherent attribute of living things upon which the preservation of the race depends. The preservation of the individual is of less importance in the scheme of Nature than the preservation of the race, and we find that food-hunger and the risk of dangers of all kinds to the continuance of an individual life are made of no account when satisfaction of mate-hunger and the preservation and perpetuation of the race requires the sacrifice or the shortening of the life, or the permanent distortion or self-immolation of the individual. Eccentric behaviour and strange exaggeration of form and colour, as judged by the standard of preservation of the individual, are found to be explained as due to structures (nervous or other) implanted in the race by natural selection, because, and in consequence of, the fact that they tend to the satisfaction of mate-hunger, and consequently to the preservation of the race.

The fact that the male animal seeks out the female in order to mate with her leads to a competition amongst males in "courtship," both in man and in the higher and lower grades of the animal series. "Courtship" comprises many procedures. Among them are the seizing and sometimes carrying off of the female by the mate-seeking male; or else the attraction of the attention of the female by the male, and her subsequent fascination by him, followed by her responsive excitement and assent to union. Fighting, often to the death, between rival suitors not unfrequently occurs.

Any animal practising the first of these arts of courtship must have developed greater strength and size than the female, and special claws or jaws or prehensile limbs which will become emphasized and increased in size by the success of the better-endowed males, and their consequent "natural selection" as parents. This elementary and violent form of courtship is found in primitive man, and is inferred to exist amongst the higher apes. It is also seen in many mammals, and in frogs and toads, and in some of the crustacea and insects which are provided with powerful claws, jaws, or limbs.

The second set of "courtship" activities mentioned above, which are of a persuasive (often hypnotic) and non-violent nature, are more widely distributed and varied. They include a number which come under the general head of "display," whether the appeal be by sound (the voice), by odour, or by strange antics and gorgeous colour. They involve the production of the most remarkable special structures ; and by their appeal to the human sense of hearing, smell, and sight are in many cases well known and familiar to us. Following upon "display" are what may be classed as "caresses" —attempts to soothe and to subjugate the female by the sense of touch.

The third kind of activity developed in "courtship" is that of fighting—fighting to the death with other suitors. It involves the production of all those natural weapons, horns, tusks, and special claws or spurs with which male animals fight one another at the breeding season. It also involves that perfection of muscular strength, rapidity, and skill in action which have enabled one male to triumph over others, and whilst destroying or banishing his less perfect opponent to transmit his own superior qualities to his offspring. It seems that to this incessantly recurring and relentless struggle between males, in courtship for the favour of the female, more rapid and important changes and developments of animal structure and endowments are due than to the more obvious competition for food, safety from enemies, and shelter. Thus muscular power, grasping and aggressive weapons, wonderful colours, forms and patterns which catch the eye, perfumes and powers of song and arresting cries, instinctive antics and caresses, have been developed in the males and transmitted to some extent to both sexes, but predominantly to the males.

Mr. Pycraft, in his book on this subject,[1] remarks that the tremendous power of "mate-hunger" has been overlooked by a strange confusion of cause and effect.

[1] "The Courtship of Animals," Hutchinson, 1913.

Almost universally its sequel, the production of offspring, has been regarded as the dominant instinct in the higher animals, but this view has no foundation in fact. Desire, for the sake of the pleasure which its gratification affords, and not its consequences, is the only hold on life which any race possesses. And this is true both in the case of man himself and of the beasts that perish. Those whose business it is, for one reason or another, to study these emotions, know well that "mate-hunger" may be as ravenous as food-hunger, and that, with some exceptions, it is immensely more insistent in the males than in the females. But for this appetite, reproduction in many species could not take place, for the sexes often live far apart, and mates are only to be won after desperate conflict with powerful rivals no less inflamed. It is idle to speak of an equality between the sexes in this matter, either in regard to animals or in the human race. The male is dominated by the desire to gratify the sexual appetite; in the female this is modified by the stimulation of other instincts concerned with the care of offspring. Amorousness is the underlying factor which has shaped and is sustaining human society, and is no less powerful among the lower animals. Much that is considered contrary to human nature, and either outrageous or ridiculous, would be understood and wisely dealt with if knowledge of nature, including man's nature, were cultivated, and took the place of vain assertions as to "what should be," accompanied by ignorance of "what is."

An excellent sample of the more violent method of "courtship by seizure" is found in the proceedings of the northern fur-seal as described by Mr. Pycraft. The old bulls, after spending the greater part of the year in the open sea, arrive at the rocks which serve as the breeding grounds a full month before the cows arrive. The younger bulls attempt, but fail, to get a place on the rocks. The bull holding the most advantageous place—the nearest to the landing-place—starts the collecting of cows. Having seized the first arrival, he places her by his side. As the later females arrive he proceeds in the same way. He soon has "herded" more cows than he can control. He cannot be in two places at once, and in scuttling off to chastise some covetous neighbour who is eloping with one of his wives, one or more bulls on the opposite side of his harem proceed to make captures from his horde. This sort of thing goes on till all the cows have been appropriated, according to the herding and holding capacities of the bulls, leaving a crowd of envious bachelors in the background not strong enough or courageous enough to fight. Each bull is master of the situation, whether his harem consists of five cows or fifty. If a cow is restless he growls at her. If she tries to escape he fiercely bites her, and if she tries to outrun him he seizes her by the skin of the neck and tosses her back, often torn and bleeding, into the family circle. Sometimes a cow is killed by the struggle of two bulls to pull her in opposite directions, and in this way the more querulous and discontented cows are eliminated in each generation, and the peculiarly gentle and passive nature characteristic of the cow seals has been developed. For three long months the bull seal has to keep watch and ward fasting. This is a most exceptional strain and effort, for in other animals fasting is associated with absolute rest and sleep. The bull fur-seal arrives at the breeding ground fat and in fine condition; he leaves it, though triumphant, a starved and battered wreck.

The more agreeable arts of courtship are exhibited by birds in greatest variety and in more familiar examples than in any other animals. The use of odours secreted by special glands as attractions to the females is frequent in the mammals—such as the musk-deer, the musk-rat, the civet, and many common hoofed animals, such as deer, antelopes, goats, and sheep—but has not been noticed in birds, though known in butterflies and moths. It is in the use of the voice in singing and in the special display of gorgeous plumage, grown, so to speak, for the purpose at the breeding season, and in strutting, fantastic posturing, and in dancing that the male bird excels. Not all birds do all these things, and female birds do none of them as a rule.

I must break off for a moment here to warn the reader that whilst we find it difficult not to speak of these activities of the male bird and male animals generally in the same terms as we speak of such behaviours in human beings, there is yet a fundamental difference between the two cases which is apt to be lost sight of in consequence of the language used. When the musk-deer and other mammals attract the female by a scent, they have no consciousness or understanding of what they are doing. They do not as a matter of thought and intention produce their perfume any more than the birds produce their gay breeding plumage by "taking thought," or the stag his great antlers or the boar his tusks. Man is, on the contrary, in these matters, as in many others, ill-provided with natural automatically-growing mechanisms of life-saving or race-perpetuating importance. Though the behaviour of man in courtship is singularly like that of many animals, he has not inherited an automatically-produced bundle of charms to allure the other sex. He has had to think the matter out and to consciously and deliberately "make" or procure from external sources both perfumes and coloured decorations and arresting (often absurd and astounding) "costumes." The males of the most savage and primitive races of men are like the bigger apes, devoid of natural "charms"; they do not allure by sweet odours, by brilliant colours, nor by caressing musical voices. They have not these possessions as natural growths of their own bodies, and they have not yet learned—probably not yet desired—to "make" or to "procure" them. There is consequently a great gulf in kind between many of the details of animal and human courtship. We have no knowledge of how the extinct creatures between ape and man stood in this respect.

In the matter of forcible seizure the conduct of the primitive man is on precisely the same footing as that of the fur-seal. As to when he began to learn from the birds and to do consciously what they do unconsciously—no one knows. In regard to the fighting with other males—man appears at a very early period to have given up the use of his natural weapons, the teeth, and to have discovered the greater utility of sharp stones and heavy clubs, and thus to have again placed himself apart from male animals, which depend on and develop automatically their tusks, horns, and claws in consequence of their value in fighting. The great interest of the jaw of the man-like Eoanthropus from Piltdown is that it was still fitted with a large canine tooth like that of a gorilla, big enough to be useful in a fight with another Piltdowner (see p. 287). But it dwindled, and in the course of time very early man-like extinct creatures were developed who had ceased to have big canines. They made use of chipped flints instead.

This substitution by man of "extraneous" weapons, decorations, and alluring appeals to the senses in place of those "intrinsic" to the animal body is all the more interesting, since we find that such substitution is already made by a number of birds, as, for instance, the magpie and the jackdaw, who collect all sorts of bright objects. The allied bower-bird of Australia makes a "play-run" or reception-room in which he places shells and bits of bone to attract the female, and the gardener bird of New Guinea clears a space in the scrub, roughly fences it and decorates it daily with bright-coloured flowers and mushrooms, freshly gathered and placed there by him, as any human bachelor may decorate his sitting-room for the delectation of his lady friends! It is a very noteworthy fact that these birds, which use extraneous decorative objects as lures, are themselves of dull plumage, but are allied to the wonderful group of Birds of Paradise, which show the greatest variety and brilliance of intrinsic decorative plumage known among birds. The love of brilliant decoration is equally keen in both groups, and is gratified in the one case by the use of extrinsic objects, in the other by the growth of intrinsic plumage. It appears that that strangely anthropoid bird—the penguin—or rather one species of penguin, familiar to Captain Scott and his companions in the Antarctic, has a similar habit of using an extraneous object as a gift or, shall we say, an excuse for an introduction when courting. The male penguin is shown in Mr. Poynting's wonderful cinema films of the Antarctic, picking up a well-shaped stone of some size and advancing with it in his beak to the lady penguin whom he has selected for his addresses. He places the stone at her feet, and retires a pace or two watching her. It is as though he said, "I am ready to build for you a first-class nest; best stones only used, of which this is a sample." If he is fortunate she looks at the stone and then at him, and without a word waddles to his side. Without more ado she accepts his proposal, and the work of constructing the stone-built nest is rapidly pushed on.

CHAPTER XXII

COURTSHIP AND DISPLAY

THE "displays" made by male birds and by some other animals which lead to the "fascination" of the females, and apparently to a condition similar to that which is called "hypnotic" in man, are very remarkable. One is tempted to say that these "displays" are made "for the purpose" of fascinating the female. But though that would be correct in describing similar proceedings on the part of a human "gallant," it is not strictly so in the case of animals, any more than it is true that a bird grows its fine plumage "for the purpose" of attracting the female. The male bird finds itself provided with fine feathers, and has probably a brief conscious pleasure in the fact, just as it has in singing, but it has, of course, no control over the growth of its feathers, nor conscious purpose in their production. Similarly, it has no knowledge or consciousness of a purpose in the antics of "display," nor in singing its melodious song, though certainly it is gratified, and has pleasurable sensations in the instinctive performances which it finds itself going through. The great French entomologist, Fabre, who has more minutely and thoroughly studied the wonderful proceedings of insects in regard to these matters and others, such as nest building, care and provision for young, deliberately says, "Ils ne savent rien de rien"—they know nothing about anything! And that is true with only small exception about even the highest animals until we come to man. Some of the higher animals have a brief and fleeting "consciousness" of what they are doing, and some of the hairy quadrupeds nearest to man have the power of "recollecting"; that is to say, have in a small degree conscious memory, and actually do reason and make use of their memory of their own individual experience to a very small and limited degree.

It is only in man that the power of reasoning—the conscious use of memory, of deciding on this or that course of action by a conscious appeal to the record of the individual's experience inscribed in the substance of the brain—becomes a regular and constant procedure. And in the lowest races of man—as, for instance, the Australian "black fellows"—this power is much less developed than in higher races, owing to the feebleness of their memory. Just as a little child or an old man recognizes the fact that his memory is bad, so does the Australian native confess to the white man that he cannot remember, and marvels at the memory of the white man, who, he says, can see both what is behind and what is to come.

"Displays" are often made by birds which have no very brilliant colours. The ruff—a bird of agreeable but sombre plumage—spreads out a ruff of feathers which grows round his neck in the breeding season and stands in a prominent position alone on the open ground with his head facing downwards and his long beak nearly touching the ground. These birds are to be seen behaving in this way at the Zoological Gardens in London. When thus posed they have a comical appearance of being absorbed in profound thought.

Suddenly, after posing for perhaps ten minutes or more immovably in this attitude, the ruff starts into life, running in a wide circle and spreading his wings, and then as suddenly relapses into his pose, with downcast eyes and beak touching the ground. This, it appears, is all a challenge to any other ruff who ventures near him, and often results in a fight with another individual who is offended by his "swagger" and attacks him. It also is an invitation and attraction to the female or "reeve" who is on the look out for a mate.

The display of the bustard, though his feathers are only light brown and white, is a very strange and arresting performance. In ordinary circumstances his feathers are nicely smoothed down, and he looks neat and fit. But at the breeding season he behaves like Malvolio when he wore cross-garters to please his lady. He approaches two or three females who are quietly feeding, and throwing his head back and his chest forward, swelling his neck out with inspired air and reflecting his tail feathers inside out (so to speak) over his back, he makes the most extraordinary havoc of his previously neat costume. The feathers are made to stand up and reflected backwards in groups, and show their underlying white surfaces round the head, on the chest, and on the wings and back, so that he suggests the appearance of a portly old gentleman, in full evening dress, the worse for liquor, his high collar unbuttoned and flapping, his short "front" bulging and loose, whilst he maintains all the time a pompous and dignified pose strangely inconsistent with his disordered costume and hesitating gait. As he struts and poses the lady bustards, though intensely interested in his strange behaviour, make no sign, and continue pecking for food, as who should say with Beatrice, "I wonder that you will still be talking, Signior Benedick: nobody marks you." After enduring this snubbing on several occasions and doggedly continuing to display his antics, the persistent bustard reaps his reward. One among the dissembling females can no longer keep up the pretence of indifference, and suddenly runs off, inviting him to follow her! The same general scheme of play is seen in the case of the peacock, who spreads his magnificent "train" around his head and neck (not to be confused with his tail, as it often is); in the case of the turkey, bubblyjock, or gobble-cock, who struts and shows off his coloured wattles and fine feathers; in that of the domestic fowl, who raises his head and neck, crows, and has a pretty trick of scraping the ground with his wing. Many other birds perform special antics suited to the display of their special plumage. Among the most varied and remarkable are those of the Birds of Paradise, which drop through the air, hang upside down on tree twigs, and pose themselves variously (often warbling the while seductive notes) according to the particular beauties which distinguish each species. Cranes and some other birds dance in groups at the mating season—really dance, making steps and jumps with the legs and movements of the wings—in rhythm.

Reptiles do only a little in the way of display. The male newt gets a crest in the spring like the wanton lapwing of Tennyson, and a splendid orange-red colour on the belly. Male fishes often develop "display" colours at the breeding season, and it is a mistake to suppose that their eyes and brains are not sensitive to colour. We have a familiar instance in the male of our common little stickleback, who, in early summer, builds, in his native pond, his nest of fragments of weed cemented together, with a wide entrance and a back door. He then becomes brilliant blood-red on the belly (he was white before) and dark green on the back, and swims about near the nest, and has an occasional fight with a competitive neighbour, whilst hustling and shepherding any female stickleback he may meet so as to make her enter it. She enters it alone, and lays an egg, or, perhaps, two or three, and then goes out by the back-door! The male, well pleased, at once goes into the nest, fertilizes the eggs, and swims out again to get another contribution to his future family. After several females have thus deposited eggs in his nest, and he has fertilized them, he keeps guard for many days whilst the young are developing. Even when they are hatched he is in constant attendance on them, for there is danger of their being eaten—not by other males, who are as busy as he is, but by the emancipated females, who neither build the nest nor care for the young, but just lay an egg here and an egg there when invited, and pursue a selfish life of amusement and voracious feeding.

It is still doubtful how far male insects of the true six-legged group appeal to the females by colour-display, even when they are brightly coloured, or in other ways than by perfumes (which they do very generally), but among the spiders there are some kinds (not common ones) in which the males have on the front of the body one or two extraordinarily brilliant spots of colour (red, apple-green, or yellow). The male moves round the female in courtship, and poses himself in most curious attitudes, so as to exhibit the brilliant colour to her; forcing it, as it were, on her attention. In other species of spiders the male dances and circles round the female, making curious and definite antics. Some spiders also have rasp-like organs, with which they can make a kind of singing note, which appears to fascinate the other sex. The vibration of a tuning-fork will cause some spiders to dance! In most spiders the female is much larger than the male—in some cases, ten times as large—and the approach of the male to the female is a dangerous business for him, for usually after his embrace she turns on him, kills him, and eats him. This is almost a unique case amongst animals (though ancient legends tell of princesses of similar ferocity), and curiously enough is not invariable among all species of spider. In some the males and females are quite friendly. The ogre-like habit of female spiders is not so injurious a thing as it may appear. For the most nourishing food is thus afforded to the female who has to ripen her eggs, and take care of her young, whilst, if the male escapes, it appears that he is short-lived and very soon dies. This cannibal tendency is very strongly developed also in the allied group, the scorpions. Two hundred scorpions were left in a cage in the South of France, whilst the naturalist (Maupertuis) who had placed them there was obliged to go to Paris. On his return he found one large, very plump and active scorpion in the box, surrounded by legs and hard bits of the bodies of the rest. The survivor was in the position of Gilbert's ancient mariner, who said that he was "the cook and the mate, and the captain's boy and the crew of the *Nancy Bell.*" Scorpions do not perform any courtship display. The males and females are of equal size, and dance together, holding one another by their large claws, before mating and retiring into a burrow.

Cuttle-fishes, squids, and the octopus—called Cephalopods—were considered by Aristotle to be the spiders of the sea. It is curious how they not only have a superficial resemblance of form to spiders, but in some habits are like them, though the Cephalopods are molluscs allied to snails and mussels, and are quite unlike spiders in deeper structure and remote from the whole group of hard-skinned, jointed-legged animals such as crustaceans, spiders, and insects. I once had the chance to see a male octopus " displaying " to a female in one of the tanks of the aquarium at Naples. There were a male and a female already living there when we introduced from another tank a second male, which had just destroyed and fed upon a large lobster, who had himself, with no evil purpose, crushed the head of a Mediterranean turtle foolishly placed by that animal between the open fingers of the lobster's big nippers. The new arrival promptly drove the earlier tenant octopus out of the tank. He pursued his rival round and round with great rapidity until the latter leapt from the surface of the water (by a violent contraction of the mantle) and escaped into the adjacent tank. Then the triumphant intruder approached the female—floods of changing colour, reddish-brown, purple, and yellow, passing over the surface of his body—and commenced an extraordinary display with his eight long sucker-bearing arms. He made these wind into close-set flat spirals and again unwind and gracefully trail in the water, when they immediately wound up again in spiral coils. The female watched this proceeding for more than an hour, and then they embraced. I could not follow any further details, but a few days after this the female piled up a number of stones, so as to make a nest in shape like a shallow basin. We enticed the male into a net and placed him in another tank, so that he should not be able to molest the female or to devour her offspring, which he would do if he had the chance. Then the female laid her eggs—minute oval, transparent bodies, each with a long stalk and all joined on to a common branching stem : the whole resembled a head of millet seed. The female tended her eggs by continually pumping a stream of water over them, and could not be driven from them. She fought savagely and heroically in their defence. But I succeeded in enticing her into a net by aid of a toothsome crab, and then took a few—only a few—of the cherished eggs, and replaced their mother in the tank, where she at once resumed the "incubation" of her eggs. For it is an "incubation," although one in which oxygenated water, and not warmth, is the accompaniment of the sitting of the "hen." I was able to watch the development of the young within the transparent eggs, which I kept in a stream of fresh sea-water, and I published a short account of what was novel in the growth of these embryos. It had not been studied previously, nor have I seen any later account of the development of octopus. The true cuttle-fish, with the hard oblong shell sunk in the back, lays each egg in a dark leathery shell. They look like small grapes, and are left, thus protected, to their fate. They have been studied, both before I obtained octopus eggs and since, in great detail. The "squid" embeds her eggs, many together, in bunches of long fingers of colourless jelly. Only the octopus and the argonaut, among Cephalopods, are known to give maternal care and incubation to their eggs.

CHAPTER XXIII

COURTSHIP, INSTINCT AND REASON

APART from the familiar instances of male colour-decoration afforded by birds, we find that even some of the minute water-fleas inhabiting freshwater lakes and the sea, and known as Crustacea Entomostraca, put on a courting dress at the breeding season; that is to say, the males become brilliantly coloured with patches of red and blue. And among the highest mammals we find that the same colours are, in some cases, displayed by the males as a fascination to the females. This is the case with the males of some of the baboons, though not with those of the highest man-like apes, who, like the primitive "savage" man, have no decoration, no pretty seductive ways appealing to either the eye or the ear, but rely on their strength and ferocity to overawe and paralyze the female. In the male "mandrill" baboon the skin of the sides of the great snout is of a deep blue colour, whilst the nose and a tract behind it is wax-like and bright red. Not only that, but the buttocks are brilliantly coloured, a central red area passing at the sides through rich purple to pale blue. The animal, which is often to be seen in menageries, is evidently proud of this finely-coloured region of his body, and turns it to a visitor and remains quietly posed, so that it may be well seen and duly admired. The hind-quarters of other monkeys, both male and female, show a brilliant red colouring during the mating season, and the skin and hair of the face is variously coloured, so as to produce a decorative pattern (eyebrows, moustache, beard, nose, all strongly contrasted in colour) in the smaller monkeys, usually more strikingly in the males than in the females. A brilliant emerald-green patch of colour is shown in the hinder part of the body of the male in one species sometimes to be seen at Regent's Park.

The making of sounds is a capacity possessed by many animals, small and big. Often it seems to have no particular significance, but, as in the case of the "humming" of bees and flies and the "droning" of beetles, is the necessary accompaniment of the vibration of the wings. But many animals make sounds as a "call," either to other individuals of their species, irrespective of sex, or more definitely as signals and appeals to the other sex, just as the luminosity which happens to accompany certain necessary chemical activities in the bodies of the lower animals has become specialized and utilized in the glow-worm and other higher forms as a signal and appeal. The rubbing of rough surfaces against one another is developed into a "stridulating organ" which we find in crickets, locusts, scorpions, spiders, and even in marine crustacea, and it is often specialized as a sexual appeal. The mere production of sound by tapping against wood is used by the little beetle, the death-watch, as a call, and is responded to by his mate with similar tapping. Such "tapping" is developed into a remarkable rhythmic vibrating sound by the birds called woodpeckers, and has its significance in courtship. But it is chiefly by the inspiration and expiration of air over vibrating cords or membranes called "vocal organs" that animals produce distinctive and musical sounds. In most cases such animals have a more general and simple "cry," which is not necessarily a sexual appeal, but addressed to comrades generally, and also a more elaborate cry or song which is primarily used by the male as an attraction in courtship, but has in the case of many birds been inherited from original male singers by the females also. The "singing" of birds—apart from simpler cries and calls—is a sexual address, an act of courtship. It is a display of power and capacity on the part of the male, and that such is its character is shown by the competition between male birds in the endeavour to "out-sing" one another. Some birds become extraordinarily excited in these competitions, which take the place of actual fighting, the victor who silences his opponents being the winner of the female bird, who is at hand listening to the competition. Caged chaffinches are celebrated for their eagerness to compete with one another in singing. They deliver their little song alternately until one is exhausted and unable to take up his turn. He is vanquished. So excited do the birds become that it occasionally happens that one of the competitors drops down dead. The beginning and directive causes of the particular song of different kinds of birds is not understood. But it is well known that they have a great gift of imitation. Parrots, piping crows, ravens, and other such birds are familiar instances, whilst little birds such as bullfinches can be trained to whistle the melodies which human beings have invented. Even the house-sparrow, which, though allied to singing finches, never sings at all when in natural conditions, has been converted into a songster by bringing it up in company with piping bullfinches.

Other animals which cannot sing like the birds yet use their voices in courtship. The frogs and toads are no mean performers in this way, whilst cats, deer, and other large animals are "singers," of a kind, when stirred by mate-hunger. The monkeys chatter and make various vocal sounds, but the gibbons and man-like apes produce excessively loud and penetrating cries. These cries, though sometimes of fine note and repeated rhythmically (as in the gibbons and chimpanzees), have not the character of song. The beginnings of song in mankind are lost in the mist of ages. The Australian black-fellows chant and dance with rhythmic precision and a certain kind of melancholy cadence, but they never attempt to fascinate the other sex by the use of the voice (nor, so far as is known, in any other way), and, indeed, there is a vast interval between their vocal performances and the love-songs of modern civilized races. Man has not inherited singing from his animal ancestry, but has re-invented it for himself. His real knowledge and command of "music" is actually a novelty which has sprung into existence within the last few hundred years.

There is no doubt that animals of the same species are attracted to one another by smell, and that distinct species have distinct smells. Further, there is no doubt that in many cases the special smell of either sex attracts the other. But modern man has so nearly lost the sense of smell—why it is difficult to say, excepting that it is because it was not of life-saving value to him—that it is very difficult for us to estimate properly the significance of perfumes and odours. We know that the dog has what to us seems a marvellous power of tracking and recognizing by smell, and that other animals appear to be similarly endowed, though most usually we cannot perceive the smell at all which they recognize and follow. It appears that nearly all the hairy quadrupeds have distinctive odours, which they and their companions can readily recognize, secreted by certain glands in the skin placed here and there on the body, often on the legs and toes. Some of these odours, like musk and civet, we can perceive, though most have no effect on us. It seems to be an evidence of the absence of any need for man to produce "perfumes" by the action of his own structure that he has a feeble sense of smell and has so little perception of any perfumes or odours peculiar to himself that he has when civilized always made use of odorous substances (perfumes and scents) extracted from other animals and from plants for the purpose, before the days of cleanliness, of masking the unpleasant odours of putrescence pervading his body and clothing. Later, when dirt became less common, he made use of perfumes for the purpose of giving an agreeable whiff to the olfactory organs of his associates.

In insects, for instance in moths and butterflies, and no doubt in most if not all others, the sense of smell is astonishingly keen, and serves as the great guide and attraction in courtship and the appeasement of mate-hunger. A single female emperor moth was placed in a box covered with fine net in a room with an open window in a country house. In three hours a dozen males of this species had entered the room, but no other moths. In twenty-four hours there were over a hundred, all fluttering around the net-covered box in which was the female. In this and other similar experiments it was found that the odour of the female moth, though imperceptible to man, clung to the box after she was removed, and that, for some days following, the empty box was nearly as powerful an attraction to the males as when it contained the female. The antennæ which carry the olfactory sense-organs are far larger in the males than in the females, as is also the case in many other lower animals where smell is a guide to mating. A single female of the vapourer moth, which is common in the London squares and parks, has been found to attract when placed in a box in an open window in Gower Street a number of males from the neighbouring plantations; and such is the penetrating and powerful character of these odorous substances produced by female moths that in one species, in which the female is wingless and lives under water, the odour escapes through the water and attracts the males in quantities to its surface. The females then arise from the depths, and, like mermaids or the witch of the Rhine, draw the infatuated males beneath the water to love and death. In several butterflies it has been shown that the males produce sweet perfumes on the surface of the wings, which can be detected as such by man, and act as stimulants to the mate-hunger of the female butterflies, which follow the scented male in numbers. The sense of smell is thus seen to be a much more powerful guide in insects than might be supposed, and it is of equally great importance to them in other enterprises and activities of life besides those of courtship. It has also a leading importance in all the lower and lowermost animals, and is the ultimate guide (for smell and taste are not separable in such simple forms) of the motile spermatic filament in its journey to the egg cell.

I have in the course of these notes on "Courtship" more than once stated that though man shares in common with all other animals the ultimate impulse to "courtship," namely, "mate-hunger," yet that it would be a mistake to suppose that he has mechanically inherited from animal ancestors (as they do) those methods of attracting and endeavouring to fascinate the female, such as the use of gay costume, dancing and posing, beautiful singing, sweet perfume, and gentle caresses, which, at various phases of his development, he has practised. True, these methods are also practised by a variety of animals, but not by man's immediate ape-like ancestors. None of these means of courtship are inherited instincts or structures in man as they are in animals. All have been arrived at and devised by man afresh, as the result of "taking thought." And in the latest advance of civilization some of them have been to a large extent either discarded or, curiously enough, handed over to the female sex. It is the woman now who endeavours to captivate the man by a display of brave colours, clothes, plumes, and jewellery, and by exquisite dancing and gesture. Not so long ago both sexes of man practised such display, but in earliest times only the male, the woman being allowed to sport a discarded rag or a broken old necklace if she were very satisfactory and submissive in her general conduct!

I must endeavour very briefly to explain how this contrast of "instinct" with "thought, knowledge, reason, and will" must (as it seems to me) be regarded. There are three great steps in the gradual evolution of the mind. The first is the slow formation (by variation and survival of the fittest) of transmissible, and therefore inherited, mechanisms of the mind, which are of various degrees of complexity, and characterize different species and kinds of animals. These mechanisms act automatically like those of a "penny-in-the-slot machine," and are just as regularly present, and as much alike in all individuals of a species, as are the other inherited structures, such as bones, flesh, viscera, the skin and its coloured clothing of decorative feathers or hair.

Later, and added to these inherited mechanisms —often interfering with them and putting an end to them—are the mechanisms of the second step. These are mechanisms arising from individual experience; they depend on memory—the inscription on "the tablets of the mind," of the experience that this follows that. They control movement and action, usurping the privilege of the previously omnipotent inherited mechanisms or instincts. This second step in the development of mind requires an excessive quantity of brain-cells. It only makes its appearance at all in animals with large brains, and reaches a far greater development in man even than in the apes, his brain being from twice to three times the size of that of the largest living ape. This use of memory and individual experience—instead of an inherited mechanism, which is the same in every member of the species—is obviously a great advantage in the struggle for existence. There are traces of it in some of the cuttlefish and insects, but even in the fishes and reptiles among living vertebrates it is of small account, and the small brain carries on its work by good, sound, inherited mechanisms or instincts, but learns nothing, comprehends nothing! In the birds we see a little—a very little—more capacity for "learning by individual experience," and it is only in the larger and later mammals that educability, or the power of learning by individual experience, becomes of serious importance. All the larger living mammals—horse, cattle, sheep, rhinoceros, tapir—have acquired an enormous increase in the size of their brains—as much as six or eight times the volume of that of their extinct ancestors whose bones and brain cavities we find fossilized in the Tertiary strata. Man has by far the biggest brain of all these animals, and has a unique degree of educability, together with the fewest instincts or in-born hereditary mechanisms among animals. He has practically to learn by individual experience—and therefore in the form best suited to his individual requirements—a host of most important actions and behaviours which even monkeys as well as dogs and sheep and horses never have to "learn," but proceed to put in practice as soon as they are born, or, at any rate, without any preliminary process of experiment and effort. Man is the one highly "educable" animal. In consequence of his large brain and its roomy memory he can be, and is—even when a "savage"—educated. Monkeys and dogs have only small "educability" as compared with man, though more than have reptiles or fishes. Man's mind is, therefore, in this essential feature different from that of animals. The modern mammals with brains as much as eight times the bulk of their early Tertiary ancestors have, it is true, acquired "educability" and the power of storing *individual* experience as "memory," but their memory is far less extensive than that of man, and though its guidance is of great value to them it acts entirely, or nearly so, without consciousness. No doubt man's brain includes some hereditary mechanisms, but in the main it distinctively consists of nerve-mechanisms, formed by his own individual education, acting on receptive and specially educable brain matter. And the brain mechanism formed by education is of greater life-saving value than is that of the inherited instincts which meet general emergencies, but not those new and special to the individual.

The third step in the development of mind is the arrival (for one can call it by no other term) of that condition which we call "consciousness"—the power of saying to oneself "I am I," and of looking on as a detached existence not only at other existences but at one's own mental processes, feelings, and movements. With it comes thought, knowledge, reason, and will. We may speak of consciousness as invading or spreading gradually over the territory of mind. All the three steps of the growth of mind which I have distinguished can be seen following one on the other in the growth of a human child from infancy to adolescence. The second step—the development of individual mechanisms due to memory—is not in most animals, and not entirely in man, pervaded by or "within the area of" consciousness. Memory is at first "unconscious memory," and there still remains in man a capacity for forming "memory" which never (or in some matters only exceptionally) becomes illuminated by consciousness. Apparently the inherited mechanisms which we call "instincts" are never within the reach of consciousness, though, of course, the actions determined by them are. It is a difficult matter to decide how far the memory of apes, dogs, and such animals nearest to man is conscious memory. Probably very little. But it is only when memory, as well as the impression of the moment, is pervaded by consciousness that reflection, and reason and action dependent on reason, are possible.[1]

Hence it is that man in all the procedure of courtship stands apart from animals. Even the Australian has not only an educable brain, but a more or less conscious memory. He seems to be permanently, in this respect, in the condition of an ordinary European child of about five years old. Gradually in the course of the development, both of increased educability and of more and more efficient and serviceable education, man has first abandoned by slow degrees his violent ancestral methods of procuring a mate, and has, as the result of observation,

[1] I have alluded to this subject again, necessarily with some repetition in the chapter on "The Mind of Apes and of Man," p. 262.

reflection, and conscious reasoning, taken to courtship by persuasion and fascination, similar to that of the birds and other remote creatures, retaining, however, for a long period his habit of fighting with other males to establish his claim to the woman of his choice. And at last, in his later development in civilized lands, he has abandoned the more obvious arts of courtship and has taken to decorating his womankind instead of himself. He has made woman take over the habit of courtship by the fascination of colour and pose whilst he looks on in sombre clothing with thoughtful reserve. He does not any longer even rely on his strength or skill in fighting in order to scatter his rivals, but makes appeal by word to the sympathy of the desired mate and trusts to the fascination which the power, given either by superior intellectual quality or by accumulated wealth, have for her.

CHAPTER XXIV

DADDY-LONG-LEGS

IN early September, golf links and other such grass-lands swarm with a large gnat-like fly of reddish-brown body, feeble flight, and long, straggling legs. These flies are generally called "Daddy-Long-Legs," or, by the more learned, "Crane-flies." I find that they are sometimes confused with another fly of about the same size with bright reddish-brown body, which is very much less abundant and occasionally flutters around the lamps and candles in a country house when the windows are open in the evening. This second kind of fly has a formidable black-coloured sting, which it shoots out from the end of its tail when handled; it has also two pairs of wings, and is an Ichneumon-fly, one of the Hymenoptera, the order of insects to which bees, wasps, ants, and gall-flies belong. Our daddy-long-legs has no sting, though the female has a sharply pointed tail. It has only one pair of wings, and belongs to the order Diptera, or tway-wing flies, in which our house-fly and bluebottle, horse-flies, tsetse-flies, gnats, and midges of vast number and variety are classified. They none of them have tail "stings," though the tail may be elongated and pointed.

Though the two-winged flies or Diptera have only two wings well grown and of full size, the second or hinder pair of wings which other insects possess of full size, are present in them in a very much dwindled condition. Since most of our common flies are very small it is

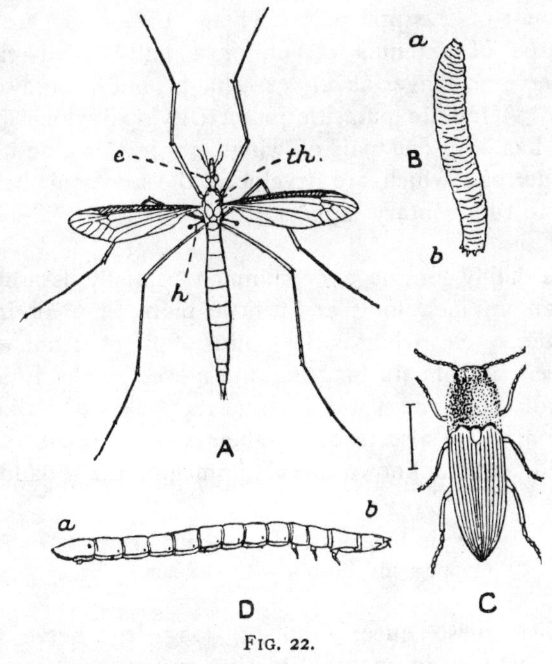

FIG. 22.

A, The Crane-fly (Daddy-Long-Legs), Tipula oleracea. *e*, the left eye; *h*, one of the balancers or "halteres," which are the modified second pair of wings; *th*, the thorax. Natural size.

B, The "Leather-jacket," the grub of the crane-fly. *a*, head; *b*, tail. Natural size.

C, The Click-beetle or Skip-jack, Elater obscurus. The line beside it shows its natural size.

D, The true Wire-worm or grub of the click-beetles. Enlarged to four times the natural length. *a*, tail; *b*, head.

difficult to see this dwindled pair of wings, which lie close behind the first or large pair, and are called the "balancers," or "halteres." The daddy-long-legs (Fig. 22, A) is big as flies go, and with a pocket lens, or even without one, you can readily see the dwindled second pair of wings standing out clearly from the body behind the attachment of the first pair. These "balancers" are of the shape of a tennis racket, or a ball-headed club. They serve no longer as organs of flight, but as auditory organs. A minute parasitic insect (Stylops) which lives in bees has only one pair of wings, but in this case it is the hinder pair which are developed, the front pair being shrunk to rudimentary lappets.

The daddy-long-legs, or common crane-fly, is a little less than an inch long and a little more than an inch across the spread wings. Its power of flight is not well developed, and its six long legs are moved so slowly and awkwardly that one would say that its powers of walking and running are also feeble. Their strange movements have led some unknown poet to imagine the "daddy" saying:

> "My six long legs, all here and there,
> Oppress my bosom with despair."

In reality these queerly-moving long legs serve the insect effectively in making its way among the closely-set blades of grass about which it crawls. The legs easily come off, and the loss of one does not appear to be a serious matter. Probably the easy detachment of a leg enables the fly to escape if one of them gets caught and nipped in overlapping blades of grass—though such a throwing away of a limb seems a rather reckless proceeding, especially since the insect has no power of "regeneration" as it is called, that is, of growing a new leg to replace the lost one. There are several well-known instances of animals which have the power of breaking off a leg or the tail if seized by an enemy or otherwise gripped. The smaller lizards and the legless lizard, called the "slow-worm," have this power in regard to the tail, but they proceed to grow a new tail after they have escaped. Some marine worms have a similar faculty, and some star-fishes (hence called "brittle-stars") have a most annoying habit of throwing off their "arms" when caught. The central disk of these star-fish, with all its arms shed, can "regenerate" the lost parts. Crabs, too, of various kinds have the habit, when caught by the leg, of breaking it off, and they may often be found with a completely-formed little leg, which has been "regenerated" or grown afresh, and will in due time attain full size. The beautiful hairy skin of the tail of the little dormouse also will come off when the animal is caught by it, leaving the bony blood-stained skeleton of the tail exposed to dry and wither up. There is no re-growth in this case. I was horrified when I was a boy to see six dormice reduced to this condition in the bird and beast shop on the staircase of the old Pantheon bazaar. They had escaped from their cage whilst I was looking on, and the shopman endeavoured to catch them, with this distressing result.

So we find that the loss of its legs by the "daddy" is a means of safety to it, and is a similar provision to that seen in some other animals. It seems improbable that the "old father long-legs" who "would not say his prayers" (according to an ancient nursery rhyme), is a myth referring to a daddy-long-legs of the insect kind, since the recommendation to "take him by his left leg and throw him downstairs" would have been futile; his left leg would have come off as soon as seized, and have greatly embarrassed the individual intending to throw him downstairs! Another kind of insect-like animal, which occurs commonly in cobwebby outhouses, and has a globular body and eight very long legs—easily broken off—is also commonly called a "daddy-long-legs." It has no wings, and is allied to the spiders, though it is not a true spider—having a minute pair of nippers near its mouth, instead of the pair of stabbing claws which spiders have. It is frequently called a "harvester," a name loosely applied to other small creatures. It is known to zoologists as Opilio.

Our crane-flies, or daddy-long-legs, when they swarm about the grass are intent on two objects. They do not require food; they have had enough when they were grubs concealed in the soil. They are now busy, first, in pairing, so that the females' eggs may be fertilized; and, secondly, the females are about to choose a likely piece of ground in which to bore with their pointed tails and lay their eggs. They prefer rather damp spots, shaded from the fierce drying heat of the sun, for this purpose. When laying her eggs, the female balances herself with her legs in an upright position, and, pushing the sharp tail into the earth, moves round by the aid of her legs, to the right and to the left, so as to bore a quarter of an inch or so into the loose soil. Then she lays two or three eggs, and, coming down from her upright pose, moves on through the blades of grass for 3 or 4 inches, and again takes an upright attitude, and repeats the boring and egglaying. The eggs are very small, black, shining grains, of which as many as 300 are found in the body of one ripe female. The male crane-fly has a broad, somewhat expanded end to its body, by which it is easily distinguished from the female.

From the eggs minute maggots or grubs hatch and feed upon animal and vegetable refuse in the soil, but as they increase in size they burrow an inch or so into the ground among the grass roots. There are two broods, one in spring and a more abundant one in August and September. The grubs have no legs. Insect grubs are often legless, as, for instance, the maggots or "gentles" of bluebottle-flies. Or they are provided with short legs, as, for instance, are the "caterpillars" or grubs of moths and butterflies. The grubs of the crane-fly (Fig. 22, B) show eleven rings or segments to the body, and have a tough grey or brownish skin, which is so firm as to give them the name of "leather-jackets." They have a head provided with a pair of short, strong mandibles or jaws, and a very short pair of feelers (antennæ). These grubs grow to be an inch and a half long, and are two-thirds the thickness of a common quill pen. They gnaw with their hard jaws the young shoots and roots of grass, and do an enormous amount of damage to grassland. They are rarely seen except when a sod is lifted, but in late spring and summer, when the grub changes to a motionless pupa or chrysalis, they may be seen protruding for about a third of their length from the surface amidst the grass tufts. Birds eat them and rooks dig with their beaks into the sod in order to pull them out, leaving a number of small pits (on the golf links) where they have been at work. The proper name of these injurious grubs is "leather-jackets." They are often confused with another grass-and-wheat pest, the "wire-worm," and are in consequence sometimes called "false wire-worms." The "wire-worm" is the grub of a beetle (Fig. 22, C and D), and is very different in appearance and history from the "leather-jacket," though both of them do great damage to grass and to grain crops.

The common crane-fly, or daddy-long-legs, is called

Tipula oleracea by entomologists, and is abundant in Europe as well as in these islands. There are other "species" of the genus Tipula common in England, namely, a smaller kind with spotted wings, Tipula maculosa, or the spotted crane-fly, and a large kind called Tipula paludosa, which frequents marsh land. There are many species of Tipula in other parts of the world, and there are closely allied kinds which are ranked in distinct genera, differing a little in certain features from the genus Tipula. These all form, taken together, the family Tipulidæ. They, together with the various kinds of gnats or "mosquitoes," the midges and fungus-flies, form one of two divisions into which the two-winged insects or Diptera are divided, namely, those with long, thread-like feelers or antennæ (Nemocera—thread horned), the other division being those with quite short antennæ (Brachycera—short horned). The latter group comprises the flies with thick, heavy bodies, such as the common house-fly, the bluebottle, the horse-flies, bott-flies, and tsetse-flies. The long-horned group have usually long, narrow bodies and long, narrow wings, which do not at once lie flat on the back when the fly alights (as do those of the short-horned group, as, for instance, those of the common house-fly). The females of the common gnat (Culex pipiens) and numerous allied species are bloodsuckers. The various midges are mostly harmless, whilst others have females which suck blood. The crane-flies do not bite. The real feeding of all these gnat-like flies is done when they are in the grub phase of their life, but the females of some gnats and midges appear to have the need of extra nourishment when in the fully-formed free-flying state, in order to ripen their large bulk of eggs. Hence, in some cases, they (but not the males) suck the juices of plants and the blood of animals.

The larval or grub phase of life is passed by many of these flies in the earth amidst putrefying vegetable and animal refuse on which they feed, as in the instance of the daddy-long-legs; but here and there we find species which penetrate into the soft parts of plants and animals. A whole group of many species burrow into mushrooms and other fungi when they are grubs; others, again, live in water when they are grubs or "larvæ," and have a very active aquatic life, rising to the surface to breathe air and searching for food in the water with their feelers and eyes, and seizing it with their powerful jaws. The mother fly in these cases lays her eggs in a group on the surface of the water or embedded in a jelly which she secretes and attaches to the leaves of water plants. Some of the short-horned flies (bott-flies and others) lay their eggs in the living flesh of warm-blooded animals, including man, and the maggots hatch there and feed on the juices of the "flyblown" animal. Cases are not rare of children being thus infested.

The black flies which fly in swarms "high" or "low" in the country lanes on summer evenings are not true biting gnats, but a large kind of midges known as Chironomus or Harlequin flies. Their eggs are laid in the water of ponds, and the larvæ on hatching bury themselves in the rich black mud and feed there. The larvæ are of a splendid blood-red colour, and are often called "blood-worms." They owe their colour to the presence in their blood of the same red oxygen-seizing crystallizable substance, hæmoglobin, which gives its red colour to the blood of man and other vertebrates. Its presence is remarkable, because in all other insects the blood is colourless or of pale blue or green tint. It seems that this hæmoglobin renders service to the larvæ of the big midges as it does to some other creatures which live in impure water, where free oxygen is very small in quantity, namely, it enables them to absorb and hold by loose chemical combination the small quantity of oxygen available. The minute midges called "Hessian fly" and "Cecidomyia"—injurious to cereal crops—should be mentioned here as among the allies of crane-flies, as also the blood-sucking midges, Ceratopogon, and the minute blood-sucking sand-flies or Buffalo-flies, called "Simulium." Species of Ceratopogon, so minute as to be barely visible, cause terrible annoyance by their bites to the salmon-fisher in Scotland, where they often swarm in countless numbers. The Buffalo-flies attack man, but in some districts of North America alight in thousands on cattle, and cause death in a few hours. A harmless long-horned fly is "the plumed fly," Corethra, the large aquatic larva of which is glass-like and quite transparent, and offers splendid facilities for microscopic research. I used to take it every year in a pond near Hampstead Heath.

The leather-jackets, or grubs of the common crane-fly (Fig. 22, B), sometimes destroy hundreds of acres—even whole districts—of grassland in England and France by gnawing the young subterranean roots and shoots of the grass. They also destroy young wheat crops. The leather-jacket is regarded by agriculturists as an intractable pest, since it gets too deep into the turf to be destroyed by chemical poisons. Its thick skin also makes it very resistant to such treatment. When immersed in brine for twenty-four hours the grubs are not killed; prolonged immersion in water is equally ineffective; they may be frozen until they are brittle, and will yet recover; and when kept three weeks without food, still remain alive. Birds are their natural enemies, and rooks not only dig after the grubs, but swallow the flies at the rate of four a minute! Ploughing up the land in which the grubs abound is recommended as a means of destroying them, and also the application of gas-lime to the ground. Rolling the turf and pressing it down also kills the grubs, but the best chance of diminishing their ravages is found in draining wet land and in feeding up the young grass plants with "fertilizers," so that they may grow rapidly and resist the injurious effect of the leather-jackets' nibbling.

Before leaving this subject it will be found interesting to contrast the "leather-jacket" with the true "wire-worms," which are the grubs of a remarkable kind of beetle (there are half a dozen British species) called the click-beetle (Fig. 22, C). They belong to a great family of beetles (Coleoptera), known as the Elaterids or Elaters, of which 7000 species are known, sixty being British. Some of the most brilliant light-giving or phosphorescent insects (not, however, the common glow-worms) belong here. The click-beetles are so called because when one is laid on its back it regains its proper pose, with the legs beneath it, by a spring or "skip," accompanied by a sharp click. The grubs of the click-beetles, known as "wire-worms" (the name is also applied to centipedes), are more threadlike, that is to say narrower, than the leather-jackets. They are not legless "maggots," but have three pairs of small legs (Fig. 22, D). They destroy corn and grass, and do not change into the adult condition in a few months, as do the leather-jackets, but remain for three, and in some cases five, years in the ground feeding on the roots of the corn and grass plants, doing much destruction before they finally change into beetles.

CHAPTER XXV

THE MOTH AND THE CANDLE

IN order to understand and interpret correctly the operation of natural selection in producing new species and maintaining them, by "the preservation of favoured races in the struggle for life" (to use Darwin's words), we must take a wide and, at the same time, a minutely accurate survey of the living world. We must seek out the evidences of this operation and use the imagination in forming conceptions as to the varied steps of the process and the results which are likely to ensue from it at different stages and in different conditions. We cannot interpret the existing structures and behaviour of living things by the use of a simple formula, such as that set up by some writers who have not properly studied Mr. Darwin's works, and declare that, according to him, all structures and behaviours which we observe in living things are perfect and the finished result of survival of the ideally fittest variations.

Plants and animals are so complex (as no one has shown more clearly than Darwin), not only in their structure but in the chemical and physical action and interaction of their living parts, that in the course of the ages during which the present species have been, step by step, fashioned in the endless vicissitudes of a changing world, many of them have retained structures or chemical constitutions which once were useful but now are useless, or even positively injurious. Even injurious structures or behaviours may be retained and inherited by a species of plant or animal, if, on the whole, the other accompanying modifications of structure are valuable—that is, of "life-saving" value, so that, "on the whole," the race is favoured by selection in the struggle for existence.

In species which have but lately acquired dominance or are brought by their success into novel conditions, we may, and do, find old structures and behaviours still persisting which are injurious, not yet, as it were, "cleaned up" and got rid of as they would be in the course of further long periods of selection. Such species become established, and may even acquire a definite stability, because the injurious structures or behaviours which they have retained are of little or no account as compared with the other advantageous characters which the species have developed. The term "disharmonies" is applied to such injurious characters, consisting in a certain want of harmony (in minor respects) between the structure of an organism and the conditions in which, nevertheless, it thrives.

Such species, imperfect because of their "disharmonies," are an illustration of the fact that Nature herself, in matters relating to living things, is not averse to compromise. Nature sets the example of toleration. Toleration may be defended on the ground that it is the biological method. Nature, though stern and inexorable as to essentials, yet accepts the faults and defects of some of her children because of the virtues and excellences which accompany them. The most highly endowed and successful forms on account of their dominance and power of spreading into new conditions, are even more likely than less highly developed kinds to retain concealed defects—disharmonies which do not lead to the destruction of the species, but occasionally cause strange embarrassment to it until they are, possibly in the long process of ages, got rid of by the slow operation of selection and survival of those individuals in which the injurious character varies in the direction of diminution and ultimate disappearance.

In man (owing, apparently, to the rapid rate at which he has been carried along towards dominance over the whole face of the globe by the development of his intelligence) the bodily structure has failed to keep pace with and to become perfected, "trimmed up," and completely adapted to, the newly-acquired habits which his increasing intelligence has forced on him. His "wisdom teeth" are "disharmonies." They are now useless and dwindled, weak spots open to the attacks of disease—since they are no longer needed for grinding coarse vegetable food, and are consequently no longer kept (by the speedy death of those individuals in whom they are small) at the full original size and efficiency seen in the apes. His large intestine is a "disharmony" not yet got rid of by natural selection, although no longer useful, but, on the contrary, the seat of poisonous putrefaction and absorption of such poisons. His tail—a few small vertebræ beneath the skin—is absolutely useless, and occasionally the seat of dangerous injury or disease. Tails very generally are liable to become useless in the descendants of animals in which they were invaluable as "fly-brushes" (cattle, horses, etc.), as prehensible organs (American monkeys), as concealing cloaks (South American ant-eater), as aids to swimming or flying, or as ornamental glories (the big-cats and others). The stumpy tail of the lynx, of some monkeys, and some lizards and fishes tells of a history in which the full-sized tail became a "disharmony"—a positive nuisance—and has been reduced, even if not abolished, by natural selection of short-tailed or tail-less varieties.

We have to be careful in asserting that any structure or behaviour in an organism is certainly a "disharmony," for it is very difficult to be quite sure as to the complete details of the life of a wild creature, and so to be able to form a conclusion rather than to suggest a possibility—as to the part played by an apparently injurious structure or habit in the economy of that creature.

One of the most striking instances of a habit or behaviour which persists and dominates the life of a wild animal to its own injury and destruction is that shown by many moths and other insects, which are attracted at night by a flame (a lamp or an open fire), and fly into it even when burnt by it, again and again until they are killed. A burnt child dreads the fire; but a burnt moth or a singed ichneumon fly seems to enjoy being burnt, and becomes more and more excited by its dashes into the flame until it finally drops with shrivelled wings to the ground. My brother told me some years ago of the verandah of his house in Java in which an open lamp was lit every night. Regularly two sets of animals, driven and guided by the action of the light on their nervous mechanism, arrived on the scene. Swarms of moths and flies dashed in and out of the flame and fell, maimed by the heat, to the ground. There a strange group had already assembled. Gigantic toads and wall lizards crept from their holes in the masonry and woodwork, and awaited the shower of injured insects, which they snapped up in eager rivalry as the infatuated flame-seekers dropped, hour after hour, to the floor. The instinct, the nervous mechanism, which brought the greedy reptiles to the spot was a "harmony," a valuable guide to nutrition; whilst the flame-seeker's impulse is assuredly a "disharmony"—a defect in adjustment—leading to death.

It is interesting to inquire into the probable origin of this fatal desire for close contact with a source of light, a desire so strong as to be entirely unchecked by the deadly heat accompanying the light. The May-flies or Ephemerids are delicate little creatures, having four net-veined wings rarely more than three-quarters of an inch across, with two or three long filaments hanging from the tail. Three hundred species are known from all parts of the world, of which forty occur in the British Islands. They live as wingless, six-legged larvæ in the water for a couple of years, feeding voraciously. Then one summer's evening they very rapidly escape from their larval skin and fly over the water in countless swarms. But only for a few hours. The eggs of the females are fertilized, and they all, both males and females, drop dead or dying into the water, where they are greedily devoured by fishes. The males are far more numerous than the females; in some species as many as 6000 males to one female have been counted. They are attracted to an extraordinary degree by lights (flames or electric lamps) set up for nocturnal illumination by civilized man, and in some districts they are collected by fishermen in this way for use as food for fish, or were so in Holland in the eighteenth century according to Swammerdam's statement in his "Biblia Naturæ." Why do they thus seek artificial lights? There is some indication of an explanation in the fact that two tropical species of May-flies are known which, like the glow-worms and fire-flies, produce light in their bodies. The May-flies, especially the males, have unusually large and prominent eyes, as is the case with phosphorescent fishes and some other light-producing animals, and it appears probable that in the now rare instances of self-luminous May-flies, the sexes are attracted to one another by the light they produce, as is the case in other luminous insects. It seems probable that the ancestral May-flies, of which many remarkable kinds have been discovered in the fossilized condition in strata as far back in time as those of the coal-measures, were all self-luminous, and acquired an overpowering instinct of seeking the light given out by other individuals as a necessary step towards sexual congress. In the course of ages other senses (probably smell and touch) have been called in to bring the fluttering insects into association. The power of producing light, being no longer needed, has disappeared from all but two rare species. But the urgent erotic instinct, the nervous mechanism, which drove the ancient May-flies towards the dancing lights of other May-flies, has remained unaltered in all the living species of the group. It is a "disharmony" which has not been of sufficient destructive importance to be "cleared away" or suppressed by natural selection. In pre-human times, nocturnal fires and lights were too uncommon to cause much disaster to the May-flies. But now that mankind sets up everywhere his nocturnal flames and electric lamps, the previously unimportant useless survival of an overpowering impulse to rush to nocturnal lights, reveals itself as a serious and death-dealing "disharmony." We must suppose, on this theory, that the other insects, such as moths and certain flies (by no means all insects), which also madly fly into nocturnal lights to their own destruction, have had luminous ancestors and a similar early history. This is a legitimate supposition, since there are several very distinct kinds of insects known at the present day which are luminous at night, although no existing moths or butterflies are known to be so.

A fact bearing on the explanation of the insects' perilous rush to flame is that birds when migrating are attracted by the great brilliant lamps of lighthouses, and, flying towards them, strike against their glass coverings, and are killed in considerable numbers. In that case, it may be that the flying towards the sun has become instinctive, and that the bright light of the lighthouse acts upon a certain number of birds (perhaps the less well-adjusted individuals) so as to call forth the same response in the direction of flight as that exercised by the sun's globe. The truth or error of this suggestion should be tested by an examination of the species of birds which kill themselves against lighthouse lanterns, and a knowledge of the season and direction of their migration.

As to luminous or phosphorescent (often called "luminescent") insects and other animals, there are a great many curious and interesting facts known. There are luminescent bacteria (common on old meat bones and dead fish and in the sea generally), animalcules of various species, jelly-fish, star-fish, worms, shell-fish, and crustaceans and true fishes. Inhabitants of the great depths of the ocean of all kinds are usually luminescent. The light is caused by the oxidation of a peculiar fatty substance. Without free oxygen there is no luminescence, and yet no heat is produced but merely light, as when a stick of damp phosphorus glows. The luminescence of living things (often, but undesirably, called phosphorescence) is a process differing greatly from that called "phosphorescence" in minerals and crystals, such as the emission of light by a lump of white loaf-sugar when crushed. You may see that kind of phosphorescence by standing in front of a looking-glass in a dark room and crushing a lump of loaf-sugar with the teeth, keeping the lips raised. It seems that in many organisms luminescence occurs without any consequent use or service to the organism. But in higher forms the power of emitting light has been seized upon by natural selection having become of value in attracting the individuals of a species to one another, or in attracting prey, or again in scaring enemies. The luminescent matter is concentrated in certain definite organs, and the access to it of oxygen and even its formation are controlled by the nervous system.

Among insects far better known than the rare luminescent May-flies, are the glow-worms, a family of beetles of which several species are known besides our own familiar one, called Lampyris noctiluca. The fire-flies of Southern Europe—Luciola italica—are small beetles allied to the glow-worm, but both sexes fly and both are luminous, whilst in the common glow-worm the female is wingless, and the flying male, who is guided to the female by her light (which she can vary in intensity), gives but a feeble light. The swarms of Italian fire-flies consist of as many as a hundred males to one female, and the males are far more brilliant than the females. My fellow-student Moseley showed some in oxygen gas at the Royal Society's soirée many years ago. The gas greatly increased their brilliancy. Many valuable experiments in search of an explanation of the brilliance of the male Luciolæ and their excess in number could be carried out in North Italy. A peculiar grub-like female glow-worm, three inches long, is found in South America, which produces a red light at each end of the body and numerous points of green light on each side of it. It is called the "railway-beetle" in Paraguay.

Another family of beetles besides the Lampyrids, or glow-worms, is celebrated for the brilliant luminescence of some of its species. These are the click-beetles, or Elaterids (see Fig. 22, C). In South America there are upwards of a hundred species of this group, showing various degrees of luminosity. The "Cucujos" (Pyrophorus noctilucus) of tropical America is one of the most abundant and largest. It is as much as an inch and a half long, and has three "lamps," or luminous organs, one on each side of the body and one below the tail. The light given off is extremely beautiful, and the live insects are used by the women for ornament and by the country-folk as lamps on nocturnal excursions. Erroneously the term "fire-fly" is applied to these beetles; it should be reserved for the little Italian Luciola, which swarms, as countless thousands of dancing lights, in the nights of early summer over the marsh lands of North Italy. I have seen it at the end of June as far north as Bonn, on the Rhine. In Australia a small true "fly"[1]—that is to say, a two-winged fly or Dipteron like our gnats, midges, and house-flies—is known, the maggot of which is luminous. And in New Zealand there is another of which both the maggot and the perfect insect are luminous. The grub is called the New Zealand glow-worm.[2]

There are grounds for believing that the luminescence of some of these insects serves them not to attract one

[1] Known to entomologists as Ceroplatus mastersi.

[2] Boletophila luminosa of entomologists.

another, but to scare would-be predatory foes, such as birds, bats, and reptiles. I have heard a story (which I should like to have confirmed) that in some part of tropical Asia a certain kind of bird collects half a dozen or so of a species of glow-worm and places them at the entrance to its nest, so as to scare nocturnal animals which might attack its eggs or its young. It is a noteworthy fact that a point of light in the dark may act in two opposite ways on animals which see it—either it attracts or it repels them. The physiologist calls this positive and negative "photo-taxis" (light-guidance). And we have the similarly positive and negative influence of chemical taste and smell, called "chemo-taxis," and a similarly contrasted positive and negative "hygrotaxis," or directive influence of moisture upon the movements of animals and plants.

CHAPTER XXVI

FROM APE TO MAN

THE recent discoveries of the actual bones of very early races of man raise again a general interest in the inquiry as to what are the actual differences of structure between men and apes, and what were probably the steps by which, as the result of "survival of the fittest," some early man-like apes became ape-like men. The question also arises as to how long ago the transition actually took place, and whether it was a very gradual or a rapid one. We are to-day in possession of some important facts bearing upon this inquiry which were unknown to Huxley when he wrote his ever-memorable essay on "Man's Place in Nature," and triumphantly closed the controversy between himself and Sir Richard Owen. That was nearly fifty years ago.

Owen had maintained that the structural difference between man and the highest apes was so great that it could only be rightly expressed by placing man in a separate sub-class of the class "Mammalia"—the hairy vertebrate animals which have warm blood and suckle their young. He pointed chiefly to the large size of the brain in man, the existence on each side of its central cavity of a little internal swelling called the "hippocampus minor," in the fanciful language of anatomists, and of the overlapping (within the skull) of the cerebellum by the hinder part of the large brain-hemispheres, or cerebrum. He called the sub-class (in which he proposed to place man alone) the "archencephala," or that of the highest developed brains (Greek, "archi," chief, and "encephalon," a brain), and proposed three other sub-classes, to contain the other orders of mammals (the Gyrencephala, Lissencephala, and Lyencephala), grouped according to three grades of complexity of the brain. Huxley denied the justification of this special grouping, by which man was placed in a separate and highest sub-class apart from the apes and monkeys. He pointed out that every bone and every part recognized by the anatomist in the higher apes is present in man (though other mammals present no such identity with him or them), and that there are only three little muscles belonging to the hand and the foot which are present in man and not present in the higher apes. He showed that the term "four-handed," or "quadrumanous," as applied to the apes and monkeys, is misleading, inasmuch as, though modified in the proportions of the digits and the mobility of the great toe, the foot of the apes has the same bones and muscles as the foot of man, and differs in structure from their hand as the foot of man differs from his hand, whilst the true hand of the apes agrees in structure with the hand of man.

Huxley (supported by many other anatomists) also showed conclusively that the little lobe in the interior of the brain called the "hippocampus minor" is present in the apes as in man, and that the posterior part of the greater brain, or "cerebrum," does overlap the "cerebellum" in apes and many monkeys to an even greater extent than it does in man. Owen's statements on this matter appear to have been due to his reliance on specimens of apes' brains removed from the skull and badly preserved in spirit—in which condition the parts in question had slipped out of their natural position. Owen's statements were thus fully demonstrated to be contrary to the fact, and Huxley declared, and conclusively showed, that so far from being entitled, on anatomical grounds, to a separate sub-class, man differs less from the higher apes—the four animals known as the gorilla, the chimpanzee, the orang-utan, and the gibbon—than does any one of these differ from the lower monkeys. Huxley came, therefore, to the conclusion that man could not logically be dissociated from the apes and monkeys in the way proposed by Owen, and that he should be placed with them in one "order," to which the name "Primates" (pronounced as three syllables, and having no reference to the clergy of the Anglican Church) is applied. This name was given by the great naturalist Linnæus, one hundred and fifty years ago, to the same group, in which, however, he erroneously included also the bats.

It was distinctly pointed out by Huxley, and has been maintained by all those who have since occupied themselves with the matter, that there are certain very obvious differences between man and the highest ape, or that which comes nearest to him in the largest number of important features—the gorilla. The chimpanzee is practically, for the purpose of such a comparison, very nearly identical with the gorilla. Both are inhabitants of tropical Africa, whilst the next nearest, the orang and the gibbons, are inhabitants of tropical Asia. The differences separating man from these near kindred animals are differences of the size and proportion of structures present in them all, and are not due to the existence in man of actual parts or structures which are present in him and not present in these apes. Man has developed from the ape, not by the production of any new organ or part, but by the definite modification of parts already present in the apes. Even that obscure internal worm-like outgrowth of the intestine, called the "vermiform appendix," which has become so unhappily familiar to the general public of late years on account of its frequent ulceration and the consequent danger to life, is present in full size in those higher apes which I have cited by name, and is present in them and man alone amongst all the varied members of the class of mammals until we come to the little Australian beaver-like "wombat," which has a vermiform appendix or narrowed tube-like extremity to the intestinal sac, called the cæcum, like that of man and the higher apes.

The changes of bodily form and proportions noticeable when we compare man with the gorilla or the chimpanzee are precisely those which fit in with the supposition of a gradual change of form and habits favoured by natural selection in the struggle for existence of ape-like creatures living originally in tropical forests, but gradually spreading beyond the special conditions of tropical life into other conditions and seeking to hold their own and to nourish themselves and their young. They have had to contend with one another for food and safety and to defend themselves either by violence or by craft against predatory animals and competitors of all kinds.

There are certain notions still current dating from Roman times as to differences between man and apes, which are simply erroneous and fanciful in origin. Thus, at one time the possession of a tail was supposed to separate animals, including monkeys and apes, from man. The rare abnormal cases in which the end of the vertebral column of man is free and projects as a tail, were, a couple of hundred years ago, cited with wonder and head-shakings as a proof that there was, after all, a real similarity in man's structure to that of animals, and pictures of the "Homo caudatus," or tailed man, were to be found in ancient books dealing with marvels and mysteries. As a matter of fact, three or four small insignificant vertebræ, almost immovable, are always present in man attached to the great bone called the sacrum, formed by the union of five vertebræ. These small vertebræ, to which the name "coccyx" is applied, are sunk beneath the skin and fat, at the end of the backbone, and though they correspond to bones of the tail of other animals, they are, in normal mankind, thus concealed from view. Precisely the same atrophy and concealment of the bones of the tail is found in the gorilla, chimpanzee, orang, and gibbons. They are all of them, seen in the flesh, as tail-less as man is, and seen in skeleton have precisely the same number of minute tail bones forming a "coccyx." This is true not only of the higher apes mentioned, but of the Barbary ape—which lives at Gibraltar—whilst others, such as the mandrill, have very short tails. In fact, the tail is a very variable appendage in monkeys, and, as the Manx breed shows, also in cats. It is mainly "decorative" in the old-world monkeys, and is probably maintained by sexual selection. It is only in the new-world monkeys that it has acquired obvious mechanical value. In them it is prehensile, and is used with great effect in swinging among the trees from branch to branch, whilst the hands and feet are left free to grasp any new support.

Another feature which is commonly, but erroneously, supposed to constitute a great difference between man and apes is the hairiness of the latter. This is only a difference of degree, for the whole surface of the body of man, excepting the eyelids, lips, palms of the hands, and soles of the feet, is covered by hair, as it is, with the same exceptions, in the apes. It is true that the hair is very fine and small on most parts of the body of man and longer on the head. But there are races of men (the Ainos of Japan and the pygmies of the Upper Nile) in which the hair on the body is coarser and more uniformly distributed than in others, and there are individuals of exceptional "hairiness" in all races of man. Moreover, before birth a coat of relatively coarse and abundant hair, called the "lanugo," is shed by the human fœtus. One variety of chimpanzee is practically bald—that is to say, has no obvious hair on the cranial region of the head. The celebrated "Sally," who lived so long in the Zoological Gardens in London, was one of this variety. When she died, I placed her brain, a remarkable one, in the museum at Oxford. Thus we see that neither tail nor hairiness separates apes from men.

So, too, the notion that animals, and therefore apes, do not and cannot laugh is erroneous. Many animals, including chimpanzees, laugh. These men-like apes also sing and dance and utter sounds (as do lower monkeys) which have definite meaning, though those sounds are very few in number and variety, and are separated by a long period of elaboration (both of skill in vocalization and in the mental development necessary to give significance to the sounds produced), from what we call "human language"—even from the speech of the most primitive of existing men.

It is often assumed as a matter of prejudice—with the intention of marking off the animal world to which the apes belong, from ourselves, the human race—that the apes show little intelligence, reasoning power, and constructive aptitudes, which might serve as the beginning of man's arts and crafts, were man derived by a slow process of development from ape-like animals of a long past geologic period. The fact is that there has been very little opportunity for studying the capacities of apes in regard to such matters, since when kept in cages they have not the opportunity of showing the skill and understanding which in their natural conditions would be obvious. The monkeys show (and this has been especially observed in the chimpanzee), in a degree greater than is seen in other animals, the mental quality which we call "curiosity." And this is combined with a persistence and determination in observation and in experiment with the purpose of satisfying that curiosity which is rarely, if ever, exhibited by other animals to anything like the same extent.

The higher apes will use their fingers to turn the screws which fasten down the lid of a box in order to see what is inside. Lately the large orang in the Zoological Gardens of London succeeded, after long efforts, in unwinding the wire fastenings of its cage and escaping into the open. It climbed into a tree, and immediately constructed for itself a platform of branches which it broke off from the tree. It then sat upon this platform, as is its habit when in its native forest. Many of the larger monkeys have great skill in throwing stones, sending them with considerable force and good aim. They select stones of size and weight appropriate to their purpose, and it would not be surprising should apes have learnt to select stones for other purposes, such as cracking nuts or the shells of molluscs, in order to extract the soft nourishing food which they contain. They are known to make use of stones for such purposes, and it would be but a short step in advance for them to choose one suitable for use as a hammer, and another suitable for use as a piercing or cutting tool. And from such a stage there is a gradual and easy passage to the simplest breaking and preparation of stones for use—in fact, to the earliest fabrication of "implements."

It is obvious when we compare not only the structure but what we know of the ways and habits of the lowest savages and the highest apes, that it is not by mere strength, swiftness, or agility that man has flourished and established himself, leaving the apes far behind him as "inferior" creatures, though as a matter of fact he is not deficient in these qualities. It is by his observation, knowledge, memory, and purposive skill that he has succeeded, and it is easy to point out a whole series of modifications of form separating man from apes, which are clearly contributory to the development of the mental qualities which give him his actual superiority. I think we are justified in taking the large opposable thumb and fingers as the starting-point in man's emergence from the ape stage of his ancestry. The exploring hand, with its thumb and forefinger, is the great instrument by which the intelligence, first of the monkey and then of man, has been developed. The thumb of the gorilla is, in proportion to the size of the fingers, very much smaller than that of man, but bigger than that of the chimpanzee, and much bigger than that of the orang and of lower monkeys. It is evident that the thumb has increased in size in the man-like apes, and in man himself this increase has been carried much further, and led to the perfecting of the hand as an instrument of exploration and construction. Contributory to the perfecting of the hand has been the gradual attainment of the upright carriage, and the use of the feet alone for walking, and the reservation of the hand for delicate exploring operations, and the bringing of objects near to the eye, to the nose, the ear, and the mouth for investigation by the great organs of special sense. The foot has become "plantigrade" in connexion with the assumption of upright carriage. It has independently become plantigrade in the gibbons and the baboons. That is to say, we and they do not walk on the edge of the half-grasping foot as do the gorilla, chimpanzee, and orang, but more steadily and firmly on its flat sole (plantar surface), as do the bears and some other animals. At the same time man has lost very greatly (but not entirely) the power of grasping with his toes. The upright carriage enabled the early ancestors of man to survey, and so to judge the conditions of safety or danger at a distance from them, as well as to devote their hands to new and special uses.

CHAPTER XXVII

THE SKELETON OF APES AND OF MAN

THE upright carriage of man has entailed remarkable changes in the proportions and shapes of parts of his body, as well as leading to special skill in the use of his hands. The vertebral column of man has not the single curve of a bow, as it has (practically) in the higher apes, but as he stands it curves (slightly, it is true, but definitely) forward at the neck, backward at the chest, forward at the loins, and backward again at the hips, an arrangement which appears to protect to some extent the brain from the transmission to it along the vertebral column of the shock caused by the sudden impact of the feet on the ground in jumping. The head is balanced on the top of this slightly elastic curvilinear column, the joint by which the skull rests on the vertebræ being placed beneath the brain-box and near the middle region of the skull. The ligaments which hold the skull in place are smaller than those in monkeys. In the higher apes the skull is not so balanced, but is held by very strong ligaments and muscles braced, as it were, on to the end of the forwardly sloping, nearly straight, backbone, from which it projects, and has further to be held in position by a great ligament attached to it and the dorsal processes of the neck vertebræ. As an adaptation to the upright carriage of man, the shape of his pelvic bones is that of a basin upon which his coiled mass of intestines can rest when he stands erect. The pelvic bones of the higher apes are flat, nearly parallel with the broad plane of the back, and give no such support; the viscera have to rest against the wall of the abdomen in the stooping position assumed by these animals in walking. The abdominal walls are consequently strong and thick, and the abdomen protrudes, as does that of a very young child. One result of man's upright carriage, showing that it is a recent acquirement and one to which he is not completely adapted, is the frequent occurrence in him of "hernia," or protrusion of the intestine through certain spaces in the deep fibrous wall of the abdomen. There would be no excessive pressure upon these spaces (near the groin), and therefore little danger of hernia, were it not for man's newly-acquired habit of erect gait. He is still incompletely adapted to the upright pose.

The arms of man are relatively shorter and his legs much longer than in the man-like apes. The Neander-men were more ape-like in these proportions than are modern races of man, and show also an "ape-like" curvature of the thigh bone which in man is straight. Whilst the arm and hand of man has gradually become a more delicate thing than that of the apes, and capable of much greater variety and efficiency in the movements of its parts, this condition has come about by alteration in proportions and to some extent shape, and not by any great change in construction. Only two muscles exist in connexion with man's hand not found in that of the higher apes. They are small slips adding to the efficiency of the fingers and thumb, whilst in the foot there is in man a small muscle connected with its outer border—"the peroneus tertius"—which helps to keep the sole of the foot turned downwards, and is not present in the apes. The general shape and proportionate size of the muscles of the leg in man give it a very different appearance from that of the ape; but there are no muscles or bones present which are not found in the apes. The beautiful outline and form of the human leg and buttocks are directly the result of the increased size of certain muscles used in maintaining the upright position, and in the peculiarly human swing of the leg in walking and running. Their beauty, like that of the other specially human features which we consider beautiful, depends upon the fact that their development, in due proportion, is a necessary condition of efficiency, activity, and strength in movements and attitudes which have gradually been acquired by man, and distinguish him from the apes. Our admiration for them is a sort of self-love, a worship of an ideal of efficiency and balance which is specifically "human," and is more or less fully realized in every individual. Probably sexual selection has had a large share in thus moulding the human form. The apes do not present the development of the gluteal region characteristic of man, and the muscles of both the arms and legs in them are, though very powerful, less fleshy and more "stringy" than those of man. There is, indeed, a difference of "quality" in the muscles of apes and men, especially civilized men, which needs investigation by the microscopist and experimental physiologist.

Though we necessarily compare man with the highest existing apes, we must not suppose that the man-like ape from which the earliest ape-like men developed was in fact a gorilla or a chimpanzee. The survival of the gorilla and chimpanzee at this day necessarily implies that they were not the actual ancestral forms which became modified and superseded in the course of man's development. Very probably the ape (the creature more ape-like than man-like, of which more anon) from which man took his direct descent had already developed a plantigrade foot—that is a foot of which the sole is placed on the ground for support, as it is in gibbons, baboons, and bears, but not in most apes, nor in cats, dogs, sheep, and horses! And probably the hands of that ancestral ape were already used more dexterously in consequence, and the dog teeth were less needed either in fighting or in breaking up food and so had become smaller.

This reflection brings us to the differences between the teeth of a man and those of apes. The face of apes is drawn forward so as to approach in form the "muzzle" of a dog. It is far less muzzle-like in the more man-like apes than in the dog-faced baboons, and in the least civilized living races of man is much less prominent—what is called "prognathous"—than in the highest existing apes. In civilized living races of man it is markedly reduced, so that in the habitual carriage of the head, with the eyes looking forward over a horizontal plane at right angles to the vertical or upright body, the front border of the jaws, in which the chisel-like incisor teeth are set, usually projects but very little beyond the brow or forehead. In Greek sculpture and other examples regarded by us as types of "beauty," the jaws do not project at all. Such a face is called "orthognathous." This modification of the shape of the face is due to the progressive dwindling in the size of the front part of the jaw and its teeth in the series dog, ape, less-civilized man, highly-civilized man, and is accompanied by an increase in the size of the front part of the brain. The number of the teeth and their arrangement in groups are identical in man and the apes. The most important difference is in the size of the front teeth, and especially in the size of the "corner" teeth (one on each side above and below), also called eye-teeth, dog-teeth, or "canines." In the highest apes, as in all monkeys, the canine teeth are very large, and even tusk-like in the males, projecting above the horizonal line formed by the crowns of the other teeth. This projecting of the canine teeth results in their not meeting one another point to point when the jaws are closed, but necessitates one, the lower, shutting in front of the other, and a space is left in the row of teeth, both in the upper and the lower jaw, for this interlocking of the great canines. It is called a "diastema." Man stands in strong contrast to the apes in this respect. His canines do not project beyond the level of the neighbouring teeth, and there is no "diastema" or gap in either the upper or lower row of his teeth.[1] There is no trace of such a gap nor any excess of size of the canines in any living race of men, and what is more remarkable, the jaws of very ancient prehistoric men which have been found in the Middle Pleistocene—the Neander or Moustierian men as well as the more ancient jaw from Heidelberg (see p. 286)—do not show any difference in this respect from the most advanced European race. On the other hand, it is one of the most remarkable features presented by the recently discovered "Piltdown" lower jaw that it had a larger canine tooth than that of any recent or fossil man, and consequently a gap or "diastema" in the row of teeth (see Chap. XXX). This difference between men and apes is all the more marked since the grinders or cheek teeth (called also molars) of man and the higher apes agree very closely, each to each in order of their position, in the pattern formed by the irregular surface of the crown. There are some slight differences

[1] See Plates VII. and VIII., p. 166, in "Science from an Easy Chair," Second Series, for careful drawings of the complete series of teeth in both the upper and lower jaw of Man and of an Ape.

in relative size and in the order of their "cutting" or growth, but these are trivial. The jaws of man show their derivation by gradual dwindling from the larger projecting jaws seen in apes and monkeys, in the close setting (that is to say, "crowding") of the teeth, and also in the dwindling and late "cutting" of the last tooth, in each jaw above and below, which we call the wisdom tooth. The "wisdom teeth" are in the higher races of men on their way to total disappearance. In lower races of men they are larger than in the higher, and in the man-like apes are of full size, and there is plenty of room in the jaw for them.

In the highest apes as well as the lower, the bony lower jaw slopes gradually backwards and downwards from the *palisade* of front chisel-like teeth or incisors (see Fig. 23 C, p. 277). There is no bony projection below the front teeth—in fact, no bony "chin." But in all modern races of men the front part of the semicircle arch of teeth has shrunk or "withdrawn" considerably or more than has the bony jaw in which the teeth are set. Consequently the bone projects in front of the front teeth as the bony chin (see Fig. 23 A, p. 277, and also "Science from an Easy Chair," 1910, pp. 404, 405). This is characteristic of modern races of man and occurs in no other animal. The very remarkable fact has recently been established that in the ancient species of man from the Middle and Lower Pleistocene—the Neander man and the Heidelberg man (Homo Neanderthalensis)—this extra or excessive shrinking of the dental arch (the half-circle formed by the complete row of teeth) had not taken place. Though the teeth are placed closely side by side and have the same shape as in modern man, they are a little bigger and form a larger and longer arch—more like a horse-shoe than a semicircle, and have not shrunk back so as to leave a projecting bony chin. The bony jaw *recedes* in these early races of men from the line of the front teeth as it does in the apes. *They have no chin* (see Fig. 25, p. 286).

Since we are all accustomed to regard a well-marked chin as a necessary feature of a beautiful human face, and to deplore or disapprove the receding or evanescent chin, it is not improbable that sexual selection has favoured the recession of the dental arch with the retention of the original bulk of the lower front margin of the jaw and chin, though why the chin should be thus appreciated is a matter of speculation. It is remarkable that in many of the monkeys the hair grows forward as a projecting beard on the front of the jaw, so as to resemble a chin although no chin is there. It is also the fact that some uncivilized races of men trim the beard and train it in a forward growth so as to suggest the possession of a very prominent chin, when in reality their solid chins of flesh and bone are not especially large.

It is not easy to suggest how the reduction in size of the canines and front teeth and of the length of the jaw could be of such advantage to incipient man as to lead to the survival of those individuals in which these parts were least developed, and so gradually to the crowding of the teeth, reduced in size, into a jaw of reduced length, whilst at a late stage, long after man was man and no ape, the teeth became so reduced in volume as to leave the lower margin of the lower jaw—projecting far in front of them as the "chin," the eminently human chin. The nutrition of these parts placed in the head near the brain, the great canine having so vast a fang that it reaches up to the eye-socket, whence it is called the "eye-tooth," renders it probable that there is a relation depending on nutrition and blood supply between them and that all-important organ contained in the neighbouring bony box, the brain. As the great teeth and long jaw have dwindled, the brain has increased in volume, and, what is more important, in activity.

Other neighbouring bony structures have dwindled whilst the brain has increased. The great longitudinal and transverse crests of bone seen on the skull of the gorilla may never have existed in that form of ape from which man is derived, but a tendency to such ridge-like outgrowth and to a greater thickness of the bony wall of the brain-case characterizes apes as distinguished from men, and its disappearance is one of the changes which have accompanied the expansion of the brain-case and the increased size of brain in man. Lower races of existing men have frequently thicker skulls than the higher races. The bony development of the skull in the higher apes is especially remarkable in the region just above the eye. The upper border of the orbit is greatly thickened, and projects as a bony arch overhanging the eye. But the extent of this growth, as also of crests on the skull, varies in individuals, and is much smaller in females than in males. In the young these ridges and prominences are absent. It is accordingly no very great change that they should disappear altogether in man, even were they as large in the ape-like ancestor of man, which probably they were not. But the existence of a considerable thickening and forward growth of the eyebrow region of the skull is noticed in many human skulls. It is particularly large in some skulls of Australian "black-fellows," and is still larger in and characteristic of the ancient species of men of the Moustierian period in Europe, Homo Neanderthalensis.

CHAPTER XXVIII

THE BRAIN OF APES AND OF MAN

A GREAT and undoubtedly very important difference between man and apes is the much greater size of the brain in man. This difference is most conveniently measured by filling the cavity of a skull, once occupied by the brain, with shot or other such material, and then measuring the bulk of the material required for that purpose. The unit which it is convenient to use in all such measurements is the cubic centimetre, because it is that used by scientific workers all over the world. A cubic centimetre is a cube the side of which is a centimetre long, and two and a half centimetres are equal to one inch. Moreover, if ever one is doubtful as to just how much an inch is, one has only to get hold of a halfpenny and mark off its breadth on a piece of paper. That is an inch, and two-fifths of it are a centimetre. Using, then, cubic centimetres as our units, we find that a good average European human brain is of the bulk of 1500 units. The gorilla has a slightly larger brain than the chimpanzee or the orang. Individual specimens differ a good deal. This is noteworthy as showing a tendency of this important organ to vary. One of good medium bulk measures 500 units, or a third of that of the well-developed European. The size of European human brains also varies within very large limits—about a third more and a third less—that is, from about 1000 units to nearly 2000. Idiots have abnormally small brains which are often deformed. We leave them aside for the moment. Healthy European adults have been measured with a brain of only 1000 units. Australian "black-fellows" have, it seems, in some cases a brain which measures as little as 900 units, but in others it reaches 1500. The skull of the fossil man from Pleistocene (possibly Pliocene) gravels in Java (known as Pithecanthropus) had a capacity of only 900 units.

If we suppose (as it is legitimate to do) that some specimens of the gorilla may have a brain a third larger than the average we get 670 units for the biggest gorilla brain, and if we similarly assume that the primitive human race of the Java gravel varied to the same extent—namely, by a third more or less around 900 as the normal—we find that the greatest size of the gorilla brain overlaps the smallest of the Javan Pithecanthropus, whilst the largest of that race would overlap not only the Australian but the smaller-sized brains of Europeans. Hence, if we accept, as we must, the fact that the brain of man and the man-like apes naturally varies greatly in volume in different individuals, there is no absolute gap in regard to size between the higher races of man and the apes. The difference is bridged over by the lower races of man and the exceptional individuals of apes.

A remarkable feature in regard to man's brain is its growth. Since it is contained in a bony box, which in the adult is firmly ossified and incapable of expansion, it is obvious that the brain, too, must cease growing when the bony box has closed in. In the apes this occurs at an earlier age than in man. The brain-box has its sides and roof constituted by a number of plate-like pieces of bone, which increase in area by addition to their margins, and finally meet each other and grow into one another, forming an irregular notched line of junction, which is called a "suture." The sutures themselves are often obliterated by bony deposit in mature life. In man the bony plates of the skull are separated by large membranous interspaces at birth—"the fontanelles"—and by delay in the junction of the bony pieces the expansion of the brain is permitted. About one-fourth of the cases of idiocy reported upon by medical observers are accompanied by an unusually small size of the brain-case (as small in some cases as 750 units), due to the premature closure of its bony walls at an unusually early period of growth. It, indeed, seems (though this is a suggestion rather than a demonstrated conclusion) that the increase of the size of the brain in normal men, as compared with apes, and the consequent development of increased mental capacity in man, may be directly set up by a delay in the ossification of the walls of the brain-case in man, as compared with his ape-like progenitors.

One of the most definite distinctions between present man and the higher apes is the length of time during which the period of growth—namely, "childhood"—and the subsequent adolescent stage of development is prolonged. The chimpanzee "Sally" was full-grown and adult at eight years of age. Savage races show maturity at an age which seems to Europeans astonishing—sometimes as early as the eleventh year. But even within the European area there is great variation in this matter, the Southern people maturing more rapidly than the Northern. There certainly is a tendency in modern civilization to defer the recognition of emergence from childhood, though whether the physical facts of growth and maturity of structure justify such a delay is not obvious. The history of our schools and universities and the records as to the age at which marriage takes place bear evidence of this modern increase of the duration of adolescence. In any case, whether the prolongation of the period of physical growth and development is even now still being increased, it is certain that the extension has taken place in former ages, and that the mental development of man is directly related in the first place to this increased period of growth, and in the second place to the prolongation of the period of organized "education" directed by the elder generation. The brain of the human child at four years of age may not infrequently reach as much as 1300 units in volume—more than double that of a full-grown gorilla—and it continues to increase in volume for some eight years, though it is difficult to say precisely when the interlocking of the bony pieces of the skull reaches a point when they can no longer yield to the expansion of the brain. The increase of the cavity of the skull practically ceases in childhood, and the increase in the size of the head subsequently is due to the increased size of muscles and fibrous structures on the outer surface of the brain-box. True as it is that man's brain is much larger than that of the higher apes, it is also true that the difference is far greater between the higher apes and the lower monkeys both as to the size of the brain and the complexity of the folds and furrows which mark the surface of the cerebral hemispheres. In these respects, as in every other anatomical feature, as was insisted by Huxley, there is less difference between man and the higher apes than between the higher apes and the lower monkeys, so that there is no pretext for placing man in a group apart from the apes and monkeys or for suggesting the existence of any great structural chasm between man and apes; on the contrary, their likeness in all important details of structure is very close.

The comparison of the size of the brain in various cases which has just been made is one of absolute size, leaving out of consideration the size and weight of the body and limbs. Putting aside the exceptional pygmy races of man (which there is no reason to regard as primitive), the average adult man is larger and heavier than the chimpanzee, and taller than, though not so powerful as, the orang. The gibbons are quite small —rarely 3 feet in height—but the male gorilla is, when adult, a much heavier animal than man, and often measures 5 feet 8 inches from the heel to the top of the head. Recently even larger specimens have been measured, and 6 feet 6 inches is quoted (probably an over-estimate) as the height attained by some specimens. This fact removes any difficulty about comparing the absolute size of brain in man and these apes. It also renders it unlikely that the primitive ape-men or men-apes were smaller than modern men, whilst the large size and weight of some of the earliest "shaped" flints (of Pliocene age) attributed to primitive man, make it probable that the men who used these flints were larger and more powerful, at any rate in the hands and arms, than modern races of men. Size and strength are, then, not points which offer any difficulty in the passage from ape to man.

What (it may well be asked) is the significance of man's greater brain? What was the advantage to man's ape-like progenitors in an increased volume of brain? It should be noted at once that the pattern of the "convolutions" marked out on the surface of the brain by a great series of winding "ditches" or "furrows" is based on one common plan in the group of monkeys and man —a plan differing from that seen in other groups which have a convoluted brain-surface—for instance from that seen in the carnivora (dogs, bears, and cats) and again from that seen in the ungulates (hoofed mammals). The convolutions of the brain of the higher apes have been minutely compared with those of man's brain. The two sets of convolutions agree very closely, but are less extensive in the apes and certain small tracts of convolutions present in man, are deficient in the apes, especially in the frontal region and at the hinder or occipital region. We know very little of the exact significance of each region of convolution in the brain. The existence of convolutions separated by furrows clearly enough increases the amount of surface of the brain, which consists of a grey substance called "the cortex of the brain," and is known to be a peculiar and specially active material. The mere comparison of the size and height of the frontal region in different animals and in man justifies the conclusion that an increase of this part of the brain is more especially related to increased intelligence. Further, the facts derived from observation of the consequences of disease or of mechanical injury in man have led to the conclusion that the "faculty of language" (the significant use of words, not the mere production of them as sounds) is especially connected with one of the frontal convolutions, which is feebly represented in the apes. The convolutions of the brain of lower races of men have not been very fully studied, but the brain of a Hottentot woman was long ago carefully described and illustrated, showing less complexity of the convolutions than is usual in European man, and making a distinct approach in this respect to the apes; but still possessing in fair proportion the convolutions characteristic of the human brain.

Abundance of convolutions and their increase at this or that part of the brain must, it is obvious, increase the active brain substance. But there is some evidence of a special kind as to the significance of increased bulk of the entire brain, apart from the folding of its surface. This is afforded by the brain cavities of the skulls observed in the series of vertebrate animals. The older groups—those "lower," that is farthest removed from man and the animals most like him—have in proportion to the bulk of their bodies much smaller brains than the later-developed groups. Thus fishes have smaller brains than reptiles, and these have much smaller brains than mammals. A cod-fish has in proportion to its bulk of living material a smaller brain than a crocodile or a turtle, and these have a much smaller brain than a pig. Not only so, but earlier kinds of mammals than the pig have a smaller brain proportionately than that animal has, and pigs have a smaller brain in proportion to their bulk than monkeys, and monkeys (as we have seen) a smaller brain than man. This increase of size is, in general, proportionate to an increase in the variety and complexity of the control of the movements of the body and their relation to the activities of the great organs of sense, such as the eyes, and the organs of smell and hearing.

But there is something more involved in the increase of the brain than this. We now know that the brain of very many kinds of animals has been increasing in size in the later geological periods. Huge reptiles as big as elephants existed on the land surface of the globe before the hairy, warm-blooded mammals which now dominate the situation had developed in number or in size—namely, in the period of and before the chalk which geologists call the Mesozoic or secondary period, to distinguish it both from the tertiary period, when mammals were abundant and large, and from the Palæozoic or primary period, at the end of which terrestrial vertebrates first began to make their appearance. These huge reptiles — such as the Iguanodon, the Triceratops, and the Diplodocus (all to be seen in skeleton, though not in the flesh, at the Natural History Museum)—had brains of an incredibly small size, much smaller in proportion to their bulk than those of living reptiles, such as lizards and crocodiles. The same extraordinary difference of size of brain is seen when we compare the large living mammals with their equally large extinct forerunners in the early tertiary strata. The skulls and whole skeletons of great rhinoceros-like animals—some of them ancestrally related to our living rhinoceroses—are dug up in early tertiary sands and clays, which have absurdly small brains. We can take a mould of the interior of the brain cases of these extinct animals and compare them with that of the recent rhinoceros. We find that the extinct animal's brain was in many cases only one-eighth the bulk of that of its modern representative!

The same disproportion in the size of the more ancient animal's brain is found when we compare the brain of the modern horse with that of its early tertiary ancestors. The modern animal has, as a rule, a very greatly increased size of brain when compared with its Miocene forefather. In fact, it seems that the brain has had, as it were, an independent development in several lines of descent, and whilst the rest of the structure of the ancestral form has been only slightly modified in its proportions, the brain cavity and the brain within it has enormously increased. It is therefore not so exceptional a thing as it at first appears—but only an instance of a change more or less widely exhibited among later animals, as compared with their near relatives in the past—when we establish the fact that the brain of the man-like apes is much bigger than that of lower monkeys, and that the brain of man, who is so closely similar in all structural details to those apes, has attained to a bulk three times that of the ape. The vast increase in the size of the brain in recent animals, as compared with their closely related representatives of an earlier period, is a frequent and regular thing. It is possible to make a suggestion, of some plausibility, as to the meaning and value of this increased size of brain, which will be found in the next chapter.

CHAPTER XXIX

THE MIND OF APES AND OF MAN

JUST as man's brain is enormously larger than that of the ordinary monkeys, although his general make and anatomy is closely similar to theirs, so we find that the rhinoceros has an enormous brain as compared with extinct rhinoceros-like animals, the predecessors and ancestors of those now living. The extinct Titanotherium of the lower Miocene period managed to carry on its life in an efficient way and to hold its own for a considerable period with a brain which was only one-eighth the bulk of that of a modern rhinoceros, as did other animals in the past with even greater bodies and smaller brains. To get some suggestion as to the significance of this fact we must, in however incomplete a way, distinguish some of the main features of the mental processes which go on in man and animals and have their "seat" in the brain.

Descartes and other philosophers have held that there is a great difference in the mental processes of animals as compared with those of man in this, namely, that man is "conscious," that is to say, conscious of himself as "I," and, as it were, looks on at himself acting on and being acted on by surrounding existences, whilst (so it is assumed) animals have not this consciousness, but are "automata," going through all the processes of life, and even behaving more or less as man does in similar circumstances, yet without being "conscious." It is, no doubt, true that many of the complicated actions of insects are carried on without consciousness of the purpose or significance of what they are doing. Such is the storing by certain wasps of smaller insects in carefully-cut chambers, to serve as food for the wasp's young, to be hatched from an egg to be laid in the "cold-storage chamber." The mother wasp will go on doing this when she has had the hind part of her body removed and has no eggs to lay. This mechanical unreasoning behaviour in insects is without exception, so that we must accept M. Fabre's conclusion that they are, in fact, unconscious "automata." I have already referred to this subject in an earlier chapter, p. 197.

We at once place ourselves in difficulty in discussing this subject by the use of the words "conscious" and "consciousness," for, as so often happens, they are customarily applied in a vague and uncertain way to the mental activities of man—without any precise agreement as to what is meant by either of them. We are all agreed that a rational human being may go through a series of elaborate actions apparently directed by purpose and yet not be what we call "conscious," that is to say, "aware" of what he is doing. This occurs in "sleep-walking" and in "day-dreaming." And again, we know that a man may be evidently conscious during a certain period, and yet forget directly afterwards that he has been conscious and said and done certain things during that period. This often happens after "concussion of the brain." It is, as a matter of fact, uncertain whether one ought to regard the condition of a man during that obliterated or forgotten period of seeming consciousness as rightly to be described by the term "conscious." And the reason why one has this doubt is that we all recognize that consciousness without memory is really a contradiction in terms. Memory—the inscription or record in our brains of past experiences—exists without consciousness, as we all know, by observation of ourselves and our fellows. But the very essence of consciousness is memory. We cannot even be "conscious" of the experience of a single moment without being also conscious of the memory of some previous condition—however small, temporary, and incomplete the memory may be. To be conscious we must *compare* the impressions reaching the brain at this moment with the memory of those of a past moment. And in lower animals and infants beginning to be conscious, the "recollections" available or accessible to consciousness may not extend farther back than a few seconds! If the memory of past experiences of which we are aware, that is to say, which are accessible to consciousness, is large and extends over the impressions of days, weeks, and years, then the conscious man or animal is in a totally different position from that of the man or animal who has only a very short and vague memory of which he or it is conscious. Thus it may be true that an animal or an infant is "conscious" and is comparing the present with the recollection of the past and yet that the basis of comparison—the reach of memory accessible to consciousness—is so small as to be of little or no significance. Yet it is the beginning of a process which, gradually enlarging the access of consciousness to "memory," passes through a thousand degrees of increasing grasp and complexity (due to increased complexity in the microscopic connexions of the structural units, the branching nerve-corpuscles, which build up the brain) until it gives us the "consciousness" of a Shakespeare, a Newton, or a Darwin. The important fact in this consideration of what we mean by "conscious" and "consciousness" is, that memory is always for most lower animals, and during a period of growth in man, untouched by consciousness, and much of it remains so in all of us. As the dawn lights up a distant peak and then another and then a whole range and spreads to valley and plain, giving greater detail and variety as the moments pass—so does consciousness slowly invade in the course of development, whether of the individual or the species, the territory of memory. In the most man-like animals and the more ape-like men the process has not gone very far. In the highest apes consciousness is so limited in its access to memory that it is but a glimmer, a mere rudiment, of what it becomes in the modern races of mankind. We must not overlook the fact that it is only when we have to deal with men far advanced from the state of primeval savagery that the memory itself becomes rich and varied. Observation, memory, and record—the vast tradition of taboo, knowledge, custom, law, and religion not inborn in our structure but handed on by spoken or written word—are developed and increased by the very fact that the daylight of consciousness has reached the memory of them when less copious than they become in later development, and has given them life-saving value.

There is no reason to doubt that consciousness—a beginning of it—exists in such animals as dogs and monkeys. And it is equally true that man not only exists for some months after he is born without being "conscious," but for some years is so only in disconnected intervals. As a matter of fact, he is very incompletely "conscious," even when adult. He is quite unconscious of a great many of his elaborate actions. He has, moreover, an "unconscious memory"—that is to say, a memory of the existence of which he is not conscious—

which guides him to, and in, the most complicated proceedings, and astonishes him when, by some chance, it is suddenly revealed to consciousness, or is converted into "conscious memory," when he dreams. Every man finds, sooner or later, that he has stored within him a register of things, persons, and events of the existence of which he was totally unaware. The gradual development of "consciousness" in higher ape-like animals and lower men, in the course of ages, is not the unparalleled thing which one is apt, at times, to consider it to be, since we can all remember the dawning of our own consciousness and its gradual development. We can also watch its growth in that most mysterious and wonderful casket of ancestral secrets and unfathomable destiny—a human infant.

Inscrutable as is the ultimate nature of "consciousness," we may put its further consideration aside on the present occasion, since it forms no actual barrier between men-like apes and ape-like men. On the contrary, the higher apes, the lower living races of men and the children of higher races, furnish us with evidence of transition from the lower condition of automatism to the higher one of self-recognition or consciousness in its most developed form. There is, however, a leading difference in the mental organization and mental processes of various animals, including man, which is of more importance in the matter which we are considering, and is largely related to the physical measurable difference in the size of the brain. The insects of which Fabre says: "They know nothing about anything," inherit a nervous mechanism—a brain and elongated mass of nerve-cells and fibres, like our spinal cord—which works sharply and definitely like a toy-automaton. Touch this part and that movement follows; excite the sense of vision with this visible thing, and such and such a movement of the limbs or jaws or other parts ensues. The stimulation of skin, eye, ear, or nose conveys a "message" by nerves to the "brain," or centre, and immediately by other nerves an "answer" is conveyed from the brain or centre in the shape of an order to this and those muscles to contract, the appropriate nerves being set at work and exciting the related muscles to contraction. The number of possible excitations and related responsive movements thus arranged is numerically very great in many animals, but they are limited. They are inherited just as they are, and come into action as soon as the necessary growth of the parts involved is attained, without hesitation or tentative trial. They are ready made. The terms "instinct" and "instinctive" should be limited to the action of this inherited apparatus or mechanism.

All animals, including man, have more or less of such an inherited instinctive nervous apparatus. Man, or for the matter of that an animal, may be "conscious" (in the sense of being "aware") of the stimulus given to this inherited apparatus, and of its related action, or he may be "unconscious" of either. The point is that we have here the "working" of an apparatus inherited in a complete working state; it is, therefore, what we call instinctive. On the other hand, there are in higher animals, and especially in man, a vast number of actions performed which are not the outcome of an inborn ready-made nervous mechanism. On the contrary, these actions are determined by a mechanism built up in the animal during its individual existence—a mechanism which is formed by its individual experience acting on its nerve-cells, and is the outcome of observation, comparison, and, more or less, of processes which we call judgment and reasoning. The persistence of this mechanism built up by the individual, as well as its continuous elaboration and development, is what we call "memory," unconscious or conscious. It is misleading to speak of "inherited memory" or "race memory," and to apply it in any way to the inherited mechanisms of instinct; the word should be reserved in its ordinary limitation to an individual's record. This new and superior apparatus appears to require a much larger bulk of brain-substance for its elaboration than that which is sufficient for the inherited mechanisms of instinct. It works in closer response to the innumerable details of the individual case, and so must be much more complicated, and we can well believe must require a larger instrument. Obviously it is an advantage to its possessor. He (be he animal or man) is provided not with a simple response suitable for the average of incidents in his life, but has, by the "education" due to the circumstances in which his individual life is carried on, formed an ever-increasing store of special little mechanisms, giving the useful or advantageous response which he has himself discovered to be appropriate to this or that sign, sound, colour, shape, smell, touch, or what not which may assail his senses. In proportion as the brain increases in volume (especially that part of it which is called "the cortex of the hemispheres") the animal to which that brain belongs loses—gets rid of—inherited mechanisms or instincts, and becomes "educable," that is to say, capable of forming for itself new individual brain mechanisms based on memorized experience.

"Educability" is the quality which distinguishes the brain of increased size. Dogs are more "educable" than rabbits; monkeys more so than dogs; and men more so—vastly more so—than monkeys and apes. The human infant is born with a few inherited mechanisms of "instinct," such as that which causes it to find its mother's nipple and to suck it, and to cling and support its own weight as no full-grown child can do. It is singularly free from any large number of inherited "instincts," and, to its own great advantage, has, during the many years in which it is protected by its parents, to learn everything and to construct new brain mechanisms—the results of "education" of the individual. We here use the word "education" in its proper and widest sense.

Thus we get an indication of "the reason why" the modern rhinoceros has a brain eight times as big as the titanotherium's. It is more "educable." The ancestors of our modern armour-plated friend have been surviving and beating their less "educable" brothers and sisters and cousins through a vast geological lapse of time; and the brains of the survivors have always been bigger, and they have become more educable and more educated until the race has culminated in those models of "sweet reasonableness," the modern rhinoceroses! It must be confessed that this character attributed to the rhinoceros is a matter of inference and not of direct observation of that animal when under his native sky. We do not judge the survivor of a fine early Miocene family by the fury and annoyance he shows when shot at, nor by the stolid contempt with which he treats mankind at the Zoo. The same signification—"educability"—attaches to the large brain of the higher apes; and man's still larger brain means still greater educability and resulting reasonableness.

In order that natural selection and the survival of the fittest should have led to this increased size and accompanying educability of the brain, it is necessary to suppose that the individuals with the more educable brain as they appeared profited by it, that is to say, did become more educated, and so defeated their rivals, and survived and transmitted their increased size of brain little by little in succeeding generations. There is no difficulty about admitting this supposition in regard to the passage from higher ape-like creatures to later forms having a full-sized brain, such as we find in the Neanderthal man and in some Australians. But we are met here by what looks, at first sight, as a fact inconsistent with our view. The obvious increased educability and consequent increased education of lower races of man by the circumstances of their lives, places them clearly enough in a position of great advantage over the higher surviving apes. But when we compare the actual mental accomplishments of the highest civilized races of man with those of big-brained savages, we find that a large proportion of individuals in the civilized races are much farther ahead of the lower savage races than most of these are ahead of the higher apes. Newton, Shakespeare, and Darwin are in mental accomplishment farther away from an Australian black, or even a Congo negro, than these "savages" are from a gorilla or a chimpanzee. Yet the difference when we compare the size and the abundance of the convolutions of the brain of the European philosopher and the black-fellow does not seem, superficially, to be proportionate to the difference in the mental performance of the two. No minute study of the microscopic differences in the structure of the two brains has, as yet, been made, and it is probable that there is a greater difference here than in the mere shape of the brain-mass. It seems that the "educability" of the brain measured by its size is little greater in the one group of men than in the other. And it is found—so far as observation and experiment have been carried—that individual savages belonging to races showing very low mental accomplishments in their native surroundings are yet capable of being "educated" to a far higher level of mental performance, when removed in early youth from their natural conditions and subjected to the same conditions as the better-cared-for children of a civilized race, than any of them ever reach in their own communities.

Very few really satisfactory experiments have been made in this direction, but the history of the negroes in America shows that the pure, unmixed negro brain is capable of showing high mathematical power, musical gifts of the best, and moral and philosophic activities equal to those of the best, or all but the exceptionally gifted, individuals of European race. It seems that the large educable brain gained by man in a relatively early period of his development from the ape has now entered on a new phase of importance. The pressure of natural selection no longer favours an increased educability (and therefore size) of brain, but the later progress of man has depended on the actual administration by each generation to its successors of an increasingly systematized exercise of that brain; in short, it has depended on education itself, and on the gigantic new possibilities of education, which have followed from the development, first, of language, then of writing, and lastly of printing, together with the accompanying growth and development of social organization, the inter-communication of all races, and the carrying on, by means of the Great Record—the written and printed documents of humanity—of the experience and knowledge of each passing generation of men from them to the men of the present moment.

Huxley agreed with Cuvier in the opinion that the

possession of articulate speech is the grand distinctive character of man. It was no sudden acquirement, but was slowly, step by step, evolved from the significant grunts and cries of apes in the course of long ages, and corresponded in its progress with a parallel progress in mental capacities. Once attained, it led to the formation of vast educative products, namely, to oral tradition, to written and then to printed memorials and records. It is not desirable in our present state of knowledge to speculate as to whether the transitional ape-man acquired the use of fire before or after he had invented articulate speech. It probably was acquired very soon after some skill in the flaking of flints had been attained, and was of immense value, both as a defence against predatory animals and as a means of preparing food. Man probably learnt at a very early period to cover himself with clothing made from the skins of other animals, and thus to tolerate cold climates. The use of clothing was correlated with the diminution of his natural hairy covering. As to the circumstances which led to the reduction in size of his canine teeth and the diminution of the projection of his jaws, it is impossible to say more than that this was favoured by the increased skill of his hand and by the use of weapons, and probably was directly correlated with an increased growth of the brain. It is an interesting fact that very young children still exhibit the ancestral tendency to bite when angry, and that the use of the teeth as weapons of attack is more frequent among lower races with "prognathous" jaws than among Europeans.

A definite habit of the human infant, that of "crying" —the peculiar spasmodic howling of very young children —seems to be unknown in any of the apes. I do not know what ingenious reason may have been assigned for this difference. Apes laugh under the same circumstances as do men, but with less production of sound than is the case with man and the hyena. Man was, far back in his monkey-days, a social and companion-loving animal, and the fact that his laughing and his weeping are accompanied by noise is due to the desire for attention and sympathy from his friends. A great difference between man and apes is the greater power of expression of various feelings or emotions by the face, and also the greater variety and significance in man of the gestures both of the upper and the lower limbs. These again are methods of seeking for and gaining sympathy and co-operation. Though not all men and not all races in an equal degree have mobility and constantly varying expression in the face, yet it is the fact that the man-like apes which have been studied in life (the chimpanzee and orang) have even less variety and range of expression than the most unintelligent savages. Man seems to have developed in an ever-increasing degree the habit of watching and interpreting the face and of giving by it expression to his emotions and states of mind, thus establishing a ready means of producing common feeling and interest in a group of associated individuals. This seems to have led to a special appreciation of the features of the face, and so to the exercise of sexual selection, resulting in what we call "a standard of beauty" in regard to both shape and expression. It is quite possible that the reduction of the threatening canine teeth and projecting jaw may have been furthered by sexual selection when once a bite had become less effective than a blow with a sharp flint, and when persuasive sounds and gestures gained more adherents than the display of tusks by a snarl.

What I have written in this and the preceding

chapters, on the differences and likenesses between apes and man and the probable steps of the transition from ape to man, may assist the reader to form a judgment as to the importance of such remains of extinct races of men as the skeleton of the Sainte Chapelle, the Heidelberg jaw, and the Piltdown jaw and cranium lately dug up in Sussex, in helping us to further knowledge of those steps. It should be definitely noted that we have not yet found any extinct animals, definitely to be classed as apes, which come nearer to man than the chimpanzee and gorilla, although we are led to infer that such creatures existed, and that their fossil remains will probably some day be discovered. On the other hand, we have in the jaw and skull recently discovered in the gravel of Piltdown, in Sussex, evidence of a man-like creature which was in most important features more ape-like than any fossil man yet discovered.

CHAPTER XXX

THE MISSING LINK

UNTIL the discovery of the wonderful fossil jaw in the gravel of Piltdown, near Lewes in Sussex, a favourite view as to the probable relationship of man and existing apes was, that if you could trace back the pedigree of man and of the chimpanzee into remote antiquity far back in the Tertiary period—probably in the early Miocene—you would arrive at a smallish creature with, proportionately to its size, larger jaws and teeth than any modern man, yet smaller than those of the living man-like apes, and with a brain not two-thirds the size of that of the least developed of modern savages, yet larger (in proportion to its general bulk) than that of the gorilla, chimpanzee, orang, and gibbons. This hypothetical creature would represent, it was held, the common ancestor of the two great "strains" or "stocks" one of which in the course of gradual modification gave rise to our living "humanity," and various non-surviving offshoots on the way; whilst the other gave rise to the company of great apes, with their tremendous jaws and dog-teeth, their small brains, and great bony skull-crests for the attachment of huge jaw muscles.

It was insisted that the obvious and immediate suggestion when once man's descent from animal ancestry was admitted, namely, that man has taken his rise from the most man-like animals we know—the great apes—is erroneous. The public was warned that they must not jump to such a conclusion; it was too obvious, too facile. The "celebrated ape of the Darwin shape," which popular songs made familiar to a wide public, was declared to be only a remote rustic, not to say brutalized, cousin of humanity, not in the direct line happily! Our real ancestors, it was declared, were mild, intelligent little creatures, animals, it is true, but animals which hastened to separate their mixed qualities in two divergent lines of descent—(1) the intelligent, mild-mannered clan who ceased to climb trees, and walked uprightly on the soles of their feet, whilst their teeth grew smaller and smaller, and their brains grew bigger and bigger; and (2) the violent tree-climbing members of the family, who refused to stand up, and acquired bigger and bigger jaws and teeth, whilst their brains remained small, their temper morose, and their conduct violent.

Old writers before the days of Darwin had talked and written about the "missing link," though I cannot say who first used the term in reference to a creature intermediate between man and apes. Sir Charles Lyell in 1851 made use of the term in regard to extinct animals which were intermediate in structure between two existing types. A learned and able writer—the Scotch judge, Lord Monboddo—in the later half of the eighteenth century put forward a theory of the development of mankind from apes such as the orang, quite independently of any general theory of "transformism" or of the progressive development of the animal and vegetable worlds, from simple beginnings. Lord Monboddo, in the absence of any knowledge of a "missing link," or of animals intermediate between man and the

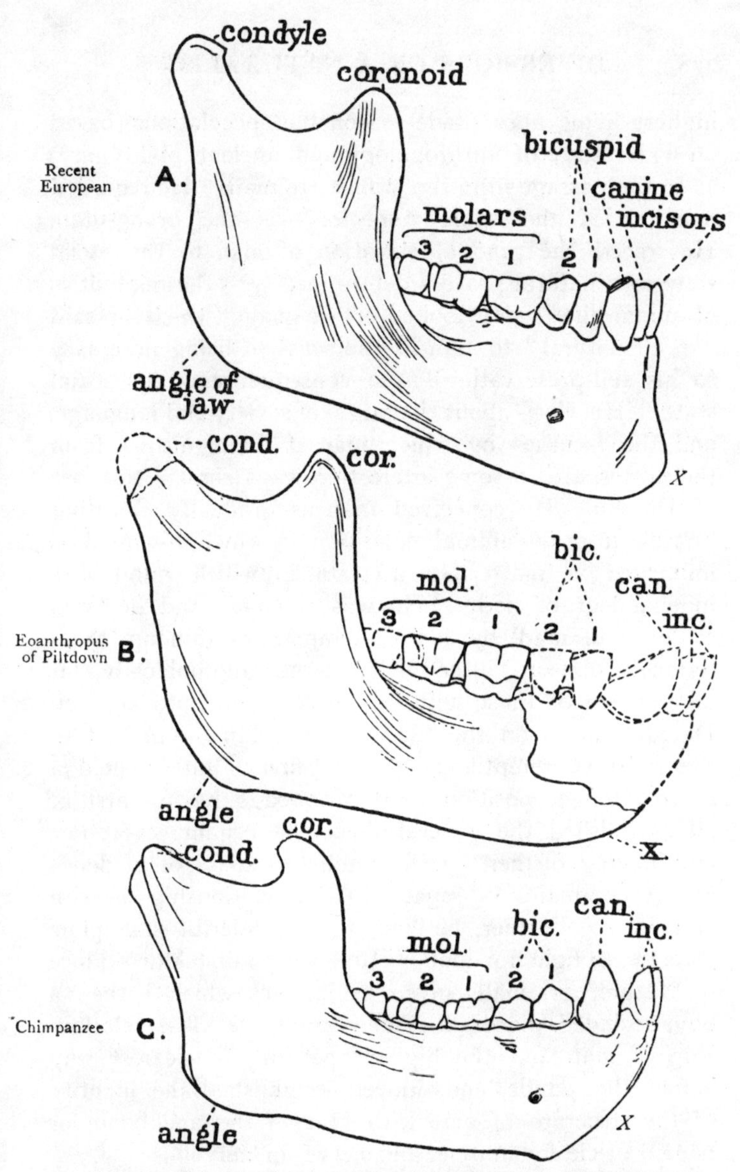

FIG. 23.—Comparison of the right half of the lower jaw of A, Modern European ; B, Eoanthropus from Piltdown ; and C, Chimpanzee. The size of the drawings is two-thirds of the linear dimensions of the actual specimens. The dotted outline in B represents the part which was wanting in the original specimen and was thus re-constructed by Dr. Smith Woodward. X in A is the bony chin or "mental protuberance" ; in B and C it marks that part of the jaw which would become the mental protuberance were the palisade or line of teeth retracted as in A.

highest living apes, made reasonable speculations (based on wide study of anthropology and ancient philosophy) as to the passage from the monkey to man. He regarded man as of the same "species" as the orang-utan. He traced the gradual elevation of man to the social state as a natural process determined by "the necessities of human life." He looked on language (which he said is not "natural" to man in the sense of being necessary to his self-preservation) as a consequence of his social state. His views about the origin of society and language, and the faculties by which man is distinguished from the brutes, are in some interesting ways similar to those of Darwin. He conceived man as gradually elevating himself from an animal condition in which his mind is immersed in matter to a state in which mind acts independent of body. He was ridiculed and declared to be half mad by his co-temporaries (among them Samuel Johnson), although he was, philosophically, far in advance of those with whom he came into contact. Darwin's views on the "Descent of Man" were met in the same contemptuous spirit at first. But he held a much stronger position than Monboddo, having first of all established the general theory of organic evolution, and having, further, a well-established mass of evidence at his command in regard to the relationship of man and apes. Further, he had that wonderful champion, Huxley, to fight for him. Huxley's book, "Man's Place in Nature," originally given as lectures which I, then a boy, attended, placed the evidence of the close relationship of man and the higher apes in the clearest way before the public, and, indeed, established the identity of the structure of man with that of the ape, bone for bone, muscle for muscle, and nerve for nerve.

Still there was always a gap—a place unfilled—

between the large-brained, small-jawed man and the small-brained, large-jawed ape. The link was missing. It was hoped, when in 1859 the human workmanship of the flint axes found with the bones of extinct animals in our river gravels was recognized, that the bones of the men who made the flint axes would turn up alongside of them, and that they would show characters intermediate between those of modern man and the great apes. But no such human bones ever were found in the older gravels deposited as terraces along their beds by the rivers of Western Europe. Human bones, and more or less complete human skulls, of a highly-developed modern type (the Cromagnards) were found in caves associated with flint tools of a different character to those common in river gravels. Then we heard a good deal about the strangely flat skull-top, or calvaria, found in a cave near Dusseldorf on the Rhine, associated with the preaching of a certain hermit named "Neuman" (= Neander). The valley was called "the Neanderthal," and the skull-top thus came to be called the "Neanderthal skull." Some authorities regarded the Neanderthal skull as that of an outcast idiot! Huxley studied it minutely, and compared it to that of Southern Australian black-fellows, and held that it took us no nearer to the apes than they did. Then an unsatisfactory small flat skull-top, together with a long, straight thigh-bone, was found in a gravel in Java, and the name "Pithecanthropus" was applied to these remains. Still we had got no nearer to any knowledge of the missing link.

Of late years we have, however, learnt a great deal more about the race or species of men of which the Neanderthal skull-top was the first indication. We now know that this species of man belonged to a period older than that of the other prehistoric cavemen—the artistic Magdalenians and the bushman-like Aurignacians, which are races of Homo sapiens, not distinct species. The older period is called the Moustierian, or Middle Paleolithic, period, and is marked by a peculiar type of flint implement. It is later than the older river gravels, in which big tongue-shaped and almond-shaped flint implements are common. The two skulls and bones from the cave of Spey, in Belgium, the Gibraltar skull, and the skeletons and skulls of the cavern called the Chapelle aux Saints in the Corrèze (Central France), and of Ferassy, and some neighbouring localities, all belong to this Moustierian age (so named after the village "Le Moustier," in Perigord), and to the peculiar species Homo Neanderthalensis.[1] It is also necessary to include here the more ancient man indicated by the important lower jaw found by Schottensack near Heidelberg (see Fig. 25). The Neanderman or Neanderthal-man had a low forehead, with overhanging bony brow-ridges, and a depressed, flattened brain-case, which, nevertheless, was very long and broad and held an unusually large brain, measuring 1600 cubic centimetres, whereas the modern European averages 1450 only of such units. He had a powerful lower jaw, with a broad, upstanding piece or vertical "ramus," and no chin protuberance. Yet his teeth were identical with those of a modern man. His thigh-bones were much curved, and his arms a good deal longer in proportion to his legs than those of a modern man. He did not carry himself upright, but with a forward stoop.

Now that we know more of him, we may ask, "Does this Neanderthal or Moustierian man fill the place of

[1] For figures of the skulls and flint implements of these ancient men, see my volume, "Science from an Easy Chair," First Series. Methuen, 1910.

the missing link?" It appears that he does not. He seems to have died out without leaving any descendants. In so far as that his bony jaw sloped directly downwards and backwards from the margin of the sockets of his front teeth, as in the apes, without projecting below, to form a chin protuberance—as it does in all races of Homo sapiens, on account of the shrinking inwards of the gum-line or palisade of front teeth (incisors and canines)—the Neanderman offers a certain approach to the condition of the apes; but in other details of shape of the lower jaw, and especially in regard to the narrowness of the lower surface of the chin and the large and deep attachments on its inner face, for the digastric muscle and certain muscles of the tongue, the bony remains of the Neanderman show that he is distinctly and altogether human, and not like the higher apes. Moreover, in the very large size of his brain (as much as 1600 units) the Neanderman shows no approach to the relatively small brain of the higher apes (which measures 500 units, possibly 800 by exception). There is in these structures some argument for the conclusion that the Neanderman could use articulate language, and inasmuch as the climate in which he flourished was extremely cold, there is ground for supposing that he could produce fire and clothe himself with skins. The flint implements which are definitely associated with him are of more skilful workmanship than the earlier, more elaborate, but less cleverly conceived, Chellean and Acheuillian implements. We cannot refuse to call him "man"—not Homo sapiens, we agree—but of the "genus" Homo—Homo Neanderthalensis.

So long as the Neanderman was the sole indication of a creature nearer in some features to the apes than are any living or extinct races of the species Homo

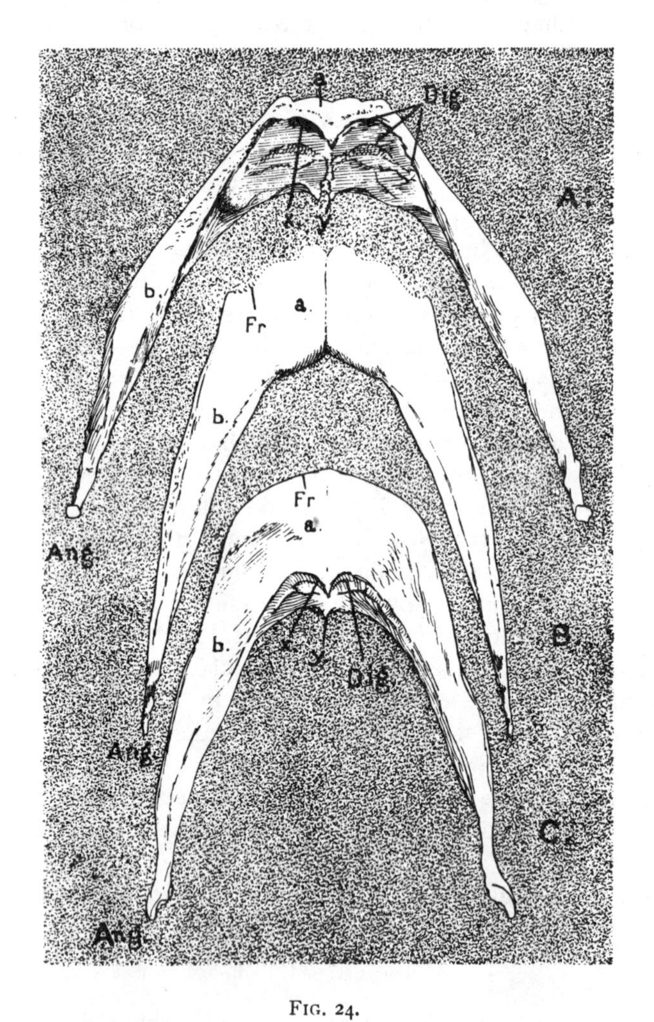

FIG. 24.

sapiens, the view was possible that the two stocks which to-day blossom and display themselves—the one as the human race, the other as the man-like apes (gorilla, chimpanzee, orang, and gibbons), became separated from one another in long past geologic ages, and that they have undergone each an independent development from a creature so unlike both as seen to-day, that we cannot speak of it as a missing link or a link at all. That view must be considerably modified by the discovery of the Piltdown jaw—the jaw of Eoanthropus Dawsoni—which is not that of a "man," that is not of the genus Homo, but must, in my judgment, be considered as one of the family Hominidæ—a Hominid, as

Description of Fig. 24.

FIG. 24.—Diagrams of the lower surface of the lower jaw of A, man; B, the Eoanthropus of Piltdown (the left half reconstructed); and C, the Chimpanzee.

The jaws are supposed to be immersed in sand, so as to conceal all but the lower surface. The narrowness of the actual inferior margin of the jaw in man, A, a, b, contrasts with the breadth and flatness of this same border in Eoanthropus, B, a, b, and the Chimpanzee, C, a, b.

In the human jaw A we see behind the narrow front border *a* the large semicircular excavations for the attachment of the digastric muscles right and left. They pass from here to the hyoid bone. From the spine (double in origin) between the two digastric impressions passes a pair of muscular slips, called the genio-hyoid muscles, also to the hyoid bone, and from the pair of spines marked *y* a pair of muscles, called the genio-glossals, pass to the tongue. These inferior and superior mental spines and the digastric impressions, much smaller in size than in man, are seen in the chimpanzee's jaw, C, but are rubbed or partly broken and partly rubbed away in the Piltdown half-jaw, B. In the figures A and C the size of the digastric impressions and mental spines is exaggerated, but their relatively much greater size in man than in the chimpanzee is correctly given, and this greater size is connected with the greater control of the tongue and the floor of the mouth in man, possibly connected with speech.

Reference Letters.—a, Broad, upwardly and forwardly sloping surface, reduced in man; b, lower border of the jaw-bone; x, front margin of the digastric "impression" of the right side. Dig, digastric impression; y, superior mental spine of the left side; Fr., fractured edge of the Piltdown jaw, and corresponding region in that of the chimpanzee.

we may say—a species assigned to a new genus Eoanthropus by Smith Woodward, which is grouped with the genus Homo and the ill-defined genus Pithecanthropus, to form the family Hominidæ; just as the genera Gorilla, Anthropopithecus (chimpanzee), Simia (orang), and Hylobates (gibbon) are grouped together to form the family Simiidæ. In Eoanthropus we have in our hands, at last, the much-talked-of "missing link"—the link obviously connecting man, the genus Homo, with the apes.

The immense importance of the discovery of the jaw of Eoanthropus by Mr. Dawson, and of the clear perception of its distinctive features by Dr. Smith Woodward, is not, as yet, sufficiently recognized. The Piltdown jaw is the most startling and significant fossil bone that has ever been brought to light. The Neandermen and the Java skull-top are simply commonplace and insignificant in comparison with it. "What leads you to say that?" I may be asked. I say so because this jaw and the incomplete skull found with it (Fig. 29) really and in simple fact furnish a link—a form intermediate between the man and the ape. Some fragments of the brain-case were found close to the jaw, indicating a fairly round, very thick-walled brain-case, holding a brain of about 1100 units capacity—very small for a man, very large for an ape. It is in the highest degree probable that the brain-case and the jaw belong to the same individual. If we were to put the brain-case aside as not certainly belonging to the same individual, we should guess that the owner of the jaw might have had a brain of about this size—intermediate between that of the larger apes and the living races of men.[1]

[1] The recent discovery by Mr. Dawson of fragments of a second skull of the same character as the first and at the same spot justifies a certain amount of hesitation in concluding that the lower jaw and the fragments of the first found skull belong to one individual.

The astonishing thing about this half-jaw from Piltdown is that it is definitely and obviously more like that of a chimpanzee—especially a young chimpanzee—than it is like that of a man (see Fig. 23, A, B, and C and their explanation). If it had been found under other circumstances it might quite well have been described as the jaw of a simiid—a large ape allied to the chimpanzee—with some unimportant resemblance to a human one. The front part of the bony jaw of Piltdown, instead of forming a narrow ridge below the protruding bony chin as in man, is wide and flat; there is no protruding chin. This very important fact is shown in our Fig. 24, in which the lower margin of the lower jaw of modern man, of the chimpanzee and of the Piltdown specimen are compared. The jaw ended in front in a wall of bone sloping forward and upward continuously from the flat and broad lower surface of the jaw. In this the great incisor teeth were set, as in all Simiids. In man, on the contrary, the front group of teeth is much smaller than in the apes, and the semicircle formed by the line of the gums is much smaller than the semicircular lower margin of the jaw. The semicircle of teeth in man retreats (as it were) behind the front part of the bony jaw which is left projecting far in advance of the line of teeth, forming the "chin" or "chin protuberance." The Piltdown jaw when found had only two of the cheek-teeth in place, as shown in Fig. 25. They were certainly very human in pattern and in the smoothness of their worn surfaces. But it was found impossible to fill the front part of the bony jaw with the missing teeth if they also were fashioned according to human pattern. They would in that case only reach along the jaw to a distance of an inch and three-fifths from the first molar tooth, whereas to fill the space from that tooth up to the front end of the bone in which the teeth are socketed they must be big enough to occupy a length of two inches and two-fifths (consult Fig. 25 and its explanation). Dr. Smith Woodward did not hesitate, in view of the shape of the jaw so closely like that of a chimpanzee, to postulate the

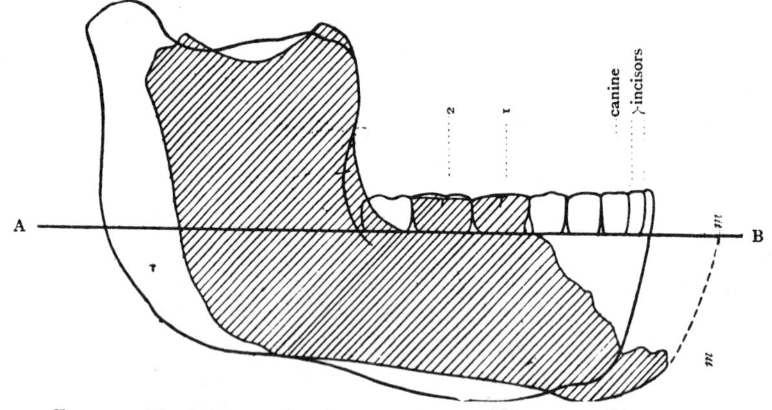

FIG. 25.—The Piltdown Jaw (shaded) and the Heidelberg Jaw (outline only) super-imposed and compared by placing the first and second molar teeth (1 and 2) of the two specimens in exact coincidence on the horizontal line A, B. The linear dimensions of the drawings are reduced to two-thirds of those of the specimens. It is obvious that when the front bony part of the Piltdown jaw is completed with an outline like that of the Heidelberg and Neander jaws, as shown by the dotted line *m*, the space between its molars and the sockets of its front teeth cannot be filled by teeth of the normal human dimensions, as it is in the Heidelberg jaw. As the figure shows, they would stop short half an inch from the front of the jaw. Hence Dr. Smith Woodward inferred that larger teeth like those of a chimpanzee were present in this region in the Piltdown jaw (Eoanthropus).

former existence in it of big front teeth—canines and incisors—like those of a chimpanzee, and unlike those of man, although there was no trace of them left in the specimen. He restored the jaw, giving it very much the shape and the teeth of a chimpanzee's jaw (Fig. 23, B). That this was a correct interpretation was proved a year later, in a startling, almost romantic way, by the discovery by Mr. Dawson and a young French naturalist who were resifting and searching the gravel at the exact spot where the jaw was found, of one of the great canine teeth, twice as big as that of any man and resembling that of a chimpanzee (see Fig. 26 and its explanation). There was a good deal of hesitation about the admission of the correctness of Dr. Smith Woodward's presentation of the jaw of Eoanthropus, with so close a resemblance to that of a chimpanzee. But the careful consideration of the specimen, and above all the welcome discovery of the great ape-like canine, has now convinced every anatomist of the truth of Dr. Woodward's restoration. The jaw itself and the recovered canine tooth, as well as the completely restored model of the two sides of the lower jaw and of the brain-case, may now be seen and studied by visitors to the Natural History Museum. They are placed in the Geological Gallery. I have visited with Mr. Dawson the gravel at Piltdown where the jaw and skull were found, and have picked

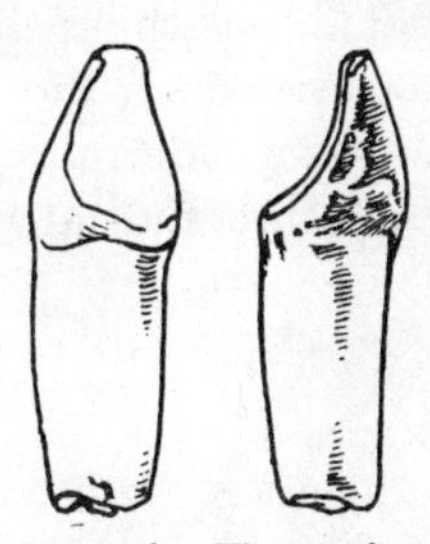

FIG. 26.—The canine tooth of the right side of the lower jaw of Eoanthropus Dawsoni, found at Piltdown a year after the discovery and description of the lower jaw, to which it belongs. Drawn of the natural size. To the left a back view, to the right a side view, showing the wearing away of the surface of the tooth.

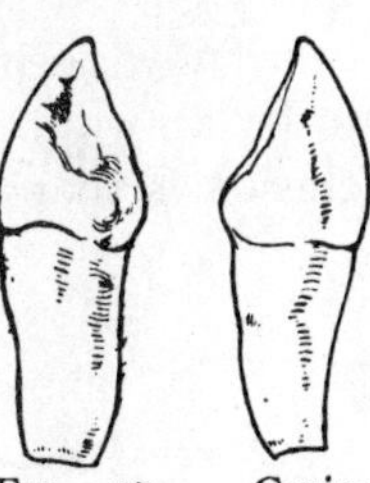

FIG. 27. — Canine tooth of the right side of the lower jaw of a European child, milk dentition. This "first" tooth is drawn of twice its actual length and breadth, which brings it very nearly to the same size as the canine of Eoanthropus. It is more closely similar in shape to the canine of the Piltdown jaw than is the canine of the second or permanent dentition of modern man.

up there humanly worked flints of very primitive workmanship. I have also followed with Dr. Smith Woodward the development and confirmation of his interpretation of the jaw.

I now desire to insist upon the legitimate conclusion to be drawn from this wonderful specimen. That conclusion is that the creature, indicated by it, is not (or was

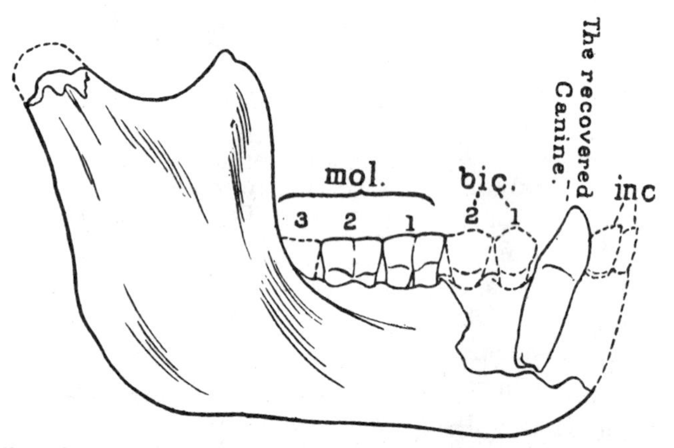

FIG. 28.—The Piltdown Jaw (Eoanthropus) with dotted lines showing the parts as now "re-constructed" or "imagined" by Dr. Smith Woodward, together with the late-found or recovered canine in its natural position.

not when it was alive) an eccentric cousin either of the Simiid or of the Hominid stock, but represents a real "missing link," an animal intermediate in great and obvious features between the two stocks, and either to be described as an ape which had become man-like or as a man who still retained characteristic ape-like features—a truly connecting or linking form. Nothing like it, nothing occupying such a position, has hitherto been discovered. It brings the focus of interest in the knowledge of primitive man away from the caves of France to the thin patch of iron-stained gravel in the meadow-land of the River Ouse as it flows through the Sussex weald. These remains are the first remains of a man-like creature found in a Pleistocene river gravel, and they exceed in interest any human remains as yet known. There is now reason to hope that more such remains will be discovered in similar gravels.[1]

It would be highly important were we able to arrive at a satisfactory conclusion as to what age must be attributed to the Piltdown jaw and skull. Did we know their age their true significance as a link between man and ape would be more easily estimated. The gravel in which they were found contains a handful, as it were, of the sweepings of the land surface of the great Weald valley of Sussex of all ages and periods since the emergence of the chalk from the ocean floor—an immense lapse of time, amounting probably to millions of years! In this sparse and inconspicuous patch of gravel we find fragments of teeth of mastodon and elephant and rhinoceros of Miocene and Pliocene age; we also find bones of quite late kinds of mammals of the Pleistocene period; we also find two kinds of roughly chipped flint instruments belonging the one to an earlier and the other to a later age. All are mixed up together in the gravel. When we come to the question as to

[1] The human lower jaw found at Moulin-Quignon fifty years ago by workmen who brought it to M. Boucher de Perthes, was dismissed after much study and examination by the most competent anatomists at the time as being a comparatively recent specimen. I do not know whether it has been preserved. I have a flint implement found with it which was given to me in 1862 by M. de Perthes as genuine. It is a forgery, and the jaw was fraudulently buried with it and others in order to deceive M. de Perthes and earn a pecuniary reward for the forgers.

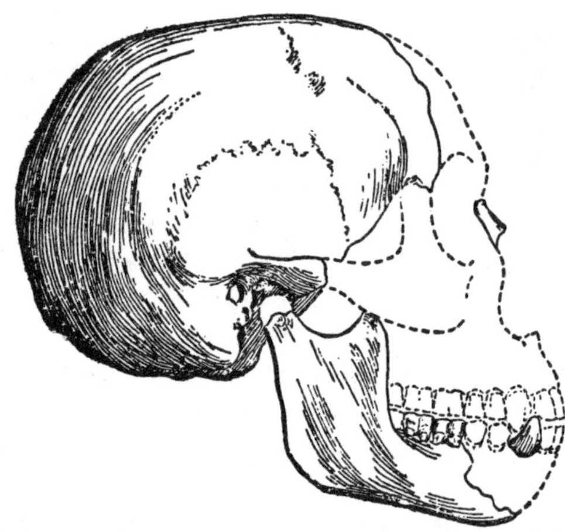

FIG. 29.—Complete Skull and Jaw of Eoanthropus Dawsoni. One-third the natural diameter. The parts indicated by dotted lines are re-constructed. The rest is drawn from the actual bones discovered at Piltdown.

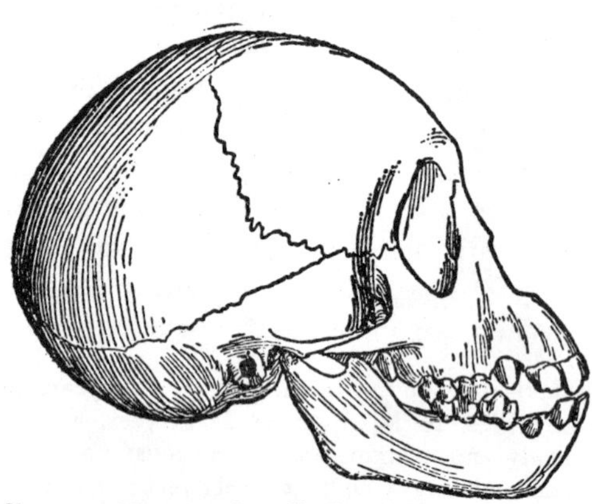

FIG. 30.—The complete Skull and Jaw of a young Chimpanzee. Drawn of one-half the natural diameter in order to compare with Fig. 29, representing the adult skull of Eoanthropus, reduced to about the same size.

which of these remains are of animals which were the contemporaries of Eoanthropus, all we can say is that Eoanthropus, the creature whose jaw was found at Piltdown, may have lived as late as the latest or as early as the earliest of the animals whose remains are associated with it. The Eoanthropus remains are not so heavily mineralized, it seems to me, as are the fragments of teeth of Miocene age found with them. At the same time, we have no ground for assuming that this creature made either the earlier or the later type of flint implements found with it, or was capable of such manufacture. I see no reason for supposing, whatever may be the age which we may have to attribute to Eoanthropus, that that creature was capable of flaking flints to a desired shape or of making fire or had developed the use of articulate speech. Nor is there any evidence to show that the humanly cut elephant-bone recently found at Piltdown by Mr. Dawson was cut by Eoanthropus. It is more probable that this was done by a more highly developed creature of the genus Homo. In fact, the only ground which at present justifies the association of Eoanthropus with the Hominidæ or human series rather than with the Simiidæ or ape series—derived from a common ancestry—is the man-like rather than ape-like size of the brain, which we must attribute to Eoanthropus on the assumption, which is at present a reasonable one, that the half-jaw and the incomplete skull found near each other at Piltdown are parts of the same individual.[1]

[1] But see foot-note on p. 284.

CHAPTER XXXI

THE SUPPLY OF PURE MILK

IT is becoming more and more certain that the character and quality of the actual things — the natural products—which we use as food and accept as "diet" are far more important matters in regard to the preservation of health than had been until recently supposed. There has been a tendency, resulting from some of the well-ascertained chemical necessities of the animal body and the equally well-ascertained chemical composition of different articles of food, to suppose that all that we have to do in regard to diet is to make sure that our food supplies us with so much carbon, hydrogen, nitrogen, and oxygen, with small quantities of phosphates, sulphates, and chlorides of potassium, sodium, calcium (lime), and iron, in a "digestible" form, in order to replace those chemical elements as their combinations are used up and thrown off as waste by our bodies. The general notions current are little more exact than this. It is recognized, it is true, that these elements must be combined in certain forms; that it is necessary to take so much "proteid" (meat, gluten of flour, casein of cheese and milk, albumen of egg), in which nitrogen is a leading component, foods which are called flesh-formers; and, further, that it is necessary to take others which supply carbon and hydrogen but have no nitrogen, namely, the hydro-carbons—fat, butter, and oil—and the carbo-hydrates—sugar and starch—foods which serve as mere fuel or heat-and-force givers. The late proprietor of "Truth," Mr. Henry Labouchere, once said to me that the doctors ought to provide us with a sausage containing in their simplest form the necessary proportions of proteid and of heat-giver (fat and sugar), and that we should abandon all "sit-down" meals, pulling the necessary sausage out of our pockets without any fuss or interruption to our occupation, and eating a couple of inches or so, three or four times a day! Experimental feeding of animals (in menageries, etc.), and even of men (in prison, on the march, and on ships), has sometimes taken very nearly as simple a form as this.

But we now know (and many, indeed, have recognized it for many years) that the nutrition of the animal body, and especially of man's body, is not so simple a matter as this method would suppose. It is necessary not merely to supply the proteids, fats, starches, and sugars, in correct weight and bulk, but also certain qualities and substances in food, much more subtle and difficult to estimate precisely, which are required in order to maintain health. There are elaborate chemical compounds present in really "fresh" meats and vegetables which seem to be absolutely necessary in order to keep man (and some of the higher animals) in health, and not only that, but it is ascertained that without them he cannot be properly nourished, but dies! These subtle, highly complex bodies seem to be present in very small quantities in good fresh food, and yet are absolutely necessary though so minute in amount. The failure of a diet consisting exclusively of tinned meats and preserved foods is due as much to this as to the nausea set up by it—of which I have written on a former occasion ("Science from an Easy Chair," Second Series, 1913, p. 171, "Food and Cookery").

Let us take an example. A distinguished medical chemist, Mr. Gowland Hopkins, has recently published an account of some experiments in which he fed young rats on a purely chemical, or "artificial" diet. He gave them, in proper proportions, chemically purified casein or curd, starch, sugar, lard, and salts, mixed into a thin paste with water, of which they had an abundant separate supply. Young rats fed with abundant natural foods of mixed substances, such as cheese, bread, egg, bits of meat and vegetable, and water, grow rapidly; they double their weight in twenty days. The young rats fed by Mr. Hopkins upon the artificial pure food—though supplied with it and taking it in abundance—did not increase in weight, and most of them died before the twentieth day! The curious and important fact was established (by careful and repeated experiment) that if a teaspoonful of milk was added to the artificial food (less than one twenty-fifth of the solid matter of their daily food) the young rats did as well as on "natural" food, doubled their weight in twenty days, and grew up to be strong and healthy rats. It was made clear that something was obtained by the rats from the small quantity of milk—something necessary for carrying on their nutrition, something the importance of which was not its quantity but its peculiar quality, which was absent in the artificial diet, but present in the mixed diet of varied materials which a young rat naturally gets. It seems that some highly elaborated proteid is necessary, if only in minute quantity, to set nutrition going, and that this is furnished by the teaspoonful of milk. Here, then, we have a case in which the simple rough conclusions as to all that is necessary in diet being the proper quantities of flesh-forming and heat-giving substance, are found to be erroneous.

Take another case—that of the disease known as "scurvy." The word "scurvy" means "afflicted with scurf, mean and dirty." It was applied to persons afflicted by this particular disease, and a Latin medical word, "Scorbutus," was made from it in the Middle Ages, which survives as "scorbutic" at the present day. Scurvy was formerly very common on board ship, in beleaguered armies, in prisons, and in other conditions in which men's food was limited to dried and salted, often badly preserved, meat and biscuit, or stale bread. Its real causation is not even now agreed upon: some holding that it was due to actual poisoning by the badly preserved food, others that it was due to the absence of certain elements—only to be obtained from fresh meat or fresh vegetables. Others think that it was caused by a bacterium. The victim of scurvy becomes much debilitated, the gums become spongy and ulcerated, and extravasations of blood are found in all parts of the body, often leading to ulceration. In the old times a whole ship's crew of the Navy would be attacked by it, and half or more died before a port could be reached and fresh food obtained. It was found that the use of fresh vegetables, fresh meat, and the juice of fruits prevented its outbreak, and cured it when once started. For one hundred and fifty years it has been held in check by the use of lime-juice as a drink whenever supplies of fresh vegetables and meat run short. It has now become so unusual a disease that there has been no proper study of it in the light of modern knowledge.

It seems to be essentially the same condition of

malnutrition as that which prevailed in cities and large tracts of country in the Middle Ages and occurs at the present day in Norway, caused by a diet of badly salted fish and dried meat. This produced ulceration of the extremities, allowing the leprosy bacillus to make its way through the broken skin into the tissues, and thus led to the widespread occurrence of leprosy. Whether bacilli of any kind were concerned in the old virulent outbreaks of "scurvy" on sailing ships must remain uncertain, but it is highly probable that they were. In any case, it is certain that the juice of fresh meat or of fresh vegetables when taken set going a better condition of nutrition in the body, and so acted as a preventive and a cure of scurvy. Some writers suppose that it was the salts, such as citrates and lactates, present in fresh fruits and vegetables which were effective in staying the disease; but this has by no means been proved, and is not, at the moment, accepted. It is probable that here, as in the case of Mr. Hopkins's rats, it was a quite minute quantity of a readily-destroyed proteid present in fresh meat and vegetables which was necessary to keep the chemical processes of nutrition in healthy activity.

This view is supported by the fact that in recent years a disease of infants similar to scurvy, and called "infantile scurvy," has been described by Sir Thomas Barlow, and fully recognized. It is a condition of "malnutrition," and is accompanied by "rickets," and is due in the first place to failure of the mother's milk, and secondly to the bad quality of the cows' milk substituted for it. Owing to the danger of infection by bovine tubercle-bacillus and the great expense of "certified" milk from specially selected cows (eightpence a quart), it is customary to boil the milk given to children.

There seems to be no doubt that good milk, freshly boiled, is satisfactory. But the constant use of sterilized milk and so-called Pasteurized milk, as well as inferior "watered" and more or less stale milk, is frequently the cause of infantile scurvy. Something is destroyed in the milk by prolonged heating which is necessary for its proper action as a food. The addition to the milk of a small quantity of fresh meat-juice or beetroot-juice appears to replace this destroyed matter, and to prevent malnutrition and scurvy. And thus the babies are rescued from "infantile scurvy." Here, again, it is a question of the presence of a minimal quantity of an easily destroyed proteid, which is necessary to start the nutritional process and to keep it going.

A very interesting case of the unsuspected influence of minute quantities of such a "proteid" body (that is, a body like casein and albumen, but higher in the complexity of its chemical structure and nearer to the readily destroyed chemical complexity of living matter itself) has lately been discovered. In the East, especially amongst Chinese "coolies" and other people who feed on rice, a very troublesome disease is known, called "Berri-berri." It is chiefly marked by pains all over the body, lassitude, and debility, and renders its victims unfit for labour, and so causes great inconvenience to employers of "Chinese cheap labour." All sorts of causes have been suggested for it. But it has now been found that it is due to the feeding of the coolies with "polished rice." This is an inferior rice, the grains of which have become (by bad, damp storage) rough and powdery on the surface. The bad rice grain is purchased by dealers and shaken up and sifted so as to get rid of this dull surface, and is then known as "polished rice." The grain has lost its outer coat. It has been found that domesticated birds (pigeons and fowls) fed on this polished rice become ill with symptoms like those of "Berri-berri," and even die. And it has been further discovered that these same birds can be cured by mixing some of the separated outer coat of sound rice grains with the "polished rice." The result of this observation on birds has been applied to human patients suffering from "Berri-berri." It is found that they are rapidly cured by giving them rice "outsides" to eat, and that those who are feeding on "polished rice" can be prevented from acquiring the disease "Berri-berri" by mixing rice "outsides" with the polished rice. The study of the subject has gone further.

A crystallizable substance allied to proteids has been separated by the chemist in quite minute quantity (one part by weight in 10,000 parts of rice) from the outer coat of rice grain, and is called "vitamine." It is this substance which prevents the "whole" rice grain from causing "Berri-berri" in men and birds who feed on it, and it has been shown experimentally that it prevents the development of "Berri-berri" when taken with "polished rice," and cures it when administered to man or bird suffering from that disease. This case calls to mind the popular notion that the indigestion caused by eating a "peeled" raw cucumber can be prevented by eating some of the dark-green "rind" or outside of the cucumber. I do not know that anyone has ever shown that this is a true doctrine, but it serves as an illustration of what has been demonstrated in the case of rice grains and "Berri-berri." Here, then, again we have, in the case of rice, a minute quantity of a substance naturally present in an article of food when taken in its natural normal condition, which is destroyed and removed by the ignorant manipulation of man, although necessary and essential if that article of food is to serve as healthy diet. In this case (as so many others) it is the attempt of greedy traders to make money by giving to a worthless spoilt article the appearance of the regular and valuable article, which has led to disease and disaster. It becomes more and more obvious that the selection of articles of food and the whole question of what is a healthy diet, are not such simple things as is often supposed. Here, as in everything we do, we must either keep to the long-established habits sanctioned by Nature, or we must have full and detailed knowledge to guide us in new ways, so that we shall not recklessly blunder by ignorance into disaster and death. The "feeding" of man and of his herds requires new and continued investigation. Old convictions and traditions in these matters must not be lightly thrust aside by the possessor of that little knowledge which is a dangerous thing. Meanwhile, for the civilized man the advice of Pasteur's pupil and successor, the late Professor Duclaux, is noteworthy: "Do not eat much, but eat many things; there is safety in variety, danger in uniformity."

When we reflect on the importance of these small quantities of easily destroyed constituents in natural foods, we begin to appreciate the difficulty of securing a pure milk-supply which shall be at the same time a nourishing and a healthy one. The sterilizing of milk by heat before it is sold as an article of diet seems to be desirable in order to destroy the bovine tubercle-bacillus which may be there and the other injurious microbes due to the dirty conditions in which the cow is kept and the milkers keep themselves. The heating of the milk for some twenty minutes to a temperature below that of boiling water seems to be the best plan. For infants, meat-juice or beet-juice may then be added to the milk when used, and so "infantile scurvy" be avoided. Consumers (older children and adults) who are taking other foods do not need this additional precaution. Milk thus "Pasteurized" is the safest milk.

But there is a very serious precaution to be observed in all cases. In such Pasteurized milk the lactic organism or ferment usually present is destroyed. Consequently the milk does not "go sour" by the growth of the lactic ferment. This is no advantage, but a serious danger. For the lactic "souring" of milk is not injurious, but, on the contrary, a safeguard. It actually prevents the growth in the milk of other really harmful and deadly germs. Thus when the lactic germ is not there, but killed by heat, these other deadly germs get their chance. A fly or other dirt-carrier brings to the sterilized milk "putrefactive" bacteria and such germs (terribly common) as those of "green" or infantile diarrhœa, not to mention others. If the milk had been unsterilized and gone sour by the growth of the lactic ferment, these more dangerous germs could not have flourished in the acid conditions produced by it. The danger of Pasteurized milk is that if kept more than a few hours at the ordinary temperature of a dwelling-room, and not carefully protected, it may be a very ready means of communicating infantile diarrhœa and other intestinal disease. It would therefore seem to be desirable to restore to the Pasteurized milk a small quantity of a pure culture of lactic germs. This could be easily done. The milk would have had its tubercle-bacilli and others removed by heat, and then, after cooling, it would receive a very few lactic germs as a protective in case it should be kept by the consumer long enough to get infected by the bacteria of intestinal disease. It is imperative that good, nourishing milk, free from germs of tubercle and of diarrhœa, shall be accessible to the millions in this country who cannot afford to pay eightpence a quart for it. It is a difficult demand to meet. What is said above explains the difficulty, and suggests an attempt to overcome it.

CHAPTER XXXII

CHRISTMAS TREES AND OTHER PINE TREES

WHEN winter grips our land it is fitting to discourse about the sweet and refreshing pine trees which are especially associated in northern climes with the celebration of Christmas. The delicious perfume which they diffuse is destructive both of microbes and noxious insects, whilst they are always linked in our minds with glorious mountain-sides or breezy moorland, or the delightful sand dunes and grey rocks of the sunny shores of the Mediterranean. The decoration of trees on days of festival and joyful celebration with garlands, lamps, and gifts is an immemorial custom of mankind, and it is probably merely the accident of its being convenient in shape, evergreen, cleanly, and sweet-smelling that has led to the selection of the common spruce as the "Christmas tree." It was not until the reign of Queen Victoria that the custom of bringing a young spruce fir into the house, growing in its special flower-pot, and then decorating it and making it the centre of a children's festival, became established in England. The 25th of December was celebrated in pre-Christian times in Northern Europe as the beginning of the New Year, and it was only after much opposition adopted by the Roman Church in the sixth century as a feast day in celebration of the birth of Christ. The Puritans rejected it as idolatrous, but its observance was restored by Charles II. In Scotland it is still ignored, and in Latin countries presents (*strenæ*, or in French *les étrennes*) are given on New Year's Day and not on Christmas Day.

The spruce is in our part of the world the commonest of the great series of cone-bearing trees which we speak of as pines and firs. Botanists call this series or "natural order" of trees the Coniferæ, in reference to the fact that their flowers are cone-shaped growths consisting of scales set in a spiral order around a central stem. Each scale is more or less overlaid by a second small scale or "bract" (sometimes evanescent), and on the inner surface of the deeper scale the naked ovules are carried in the female cones, whilst the pollen-producing growths are similarly carried by the smaller and more delicate male cones. The ovules are exposed nakedly, and are, therefore, in a more primitive condition than those of ordinary flowering plants, in which they are overgrown and enclosed by the modified leaves which form the "pistil" or central part of the flower. Hence the conifers are called flowering plants with "naked seeds," or Gymnosperms, whilst the rest of the flower-bearing plants are called plants with "covered seeds," or Angiosperms. The cones are at first green (sometimes purple), and become brown as they ripen. The small loosely-packed male cones, less familiar to most people than the solid and large seed-bearing cones, are often of a fine crimson colour when young, and when ripe of a bright chestnut brown, but the cones of pine trees are with few exceptions (the Douglas fir is one) not brilliantly coloured nor set out to attract the eye, as are the flowers of most flowering plants. Though a young branch carrying its groups of green "needles," rich brown male cones, silver-white hairs and swelling seed-cones (Fig. 31) presents a very fine harmony of diverse colours, yet they are not constructed so as to attract the visits of insects. They do not require the services of insects to carry the pollen of the male cones to the ovules of the female cones. They produce an enormous amount of pollen, which falls in showers of yellowish-white dust, and is blown by the wind, far and wide, on to the female cones. Hence it is that though the cones are "flowers," and the pine trees are flowering plants, yet they have none of the beautiful shapes and colours which we associate, as a rule, with flowers—shapes and colours due to the modification in the latter of the leaves called "petals" which are set with attractive brilliancy around the stamens and pistil. The conifers are an ancient race, dating from geological ages before the chalk, when plants had not "learnt" (as they subsequently did) to colour their flowers and to provide nectar so as to ensure the visits of insects and the carriage by them of their pollen from plant to plant. Even in the group of plants with coloured flowers there are trees which have abandoned the production of colour in their flowers, and like the conifers depend upon the wind to carry their pollen instead of seeking the aid of insects.

The word "pine" is of Latin origin, and belongs properly to the South of Europe; the word "fir" is Teutonic, and is originally applied to the same trees in the North of Europe as those to which "pine" is applied in the South. It is of no use trying to determine what conifers should rightly be called "firs" or "fir trees," and which "pines" or "pine trees." There is complete confusion and indifference nowadays in the use of those words, and the botanists have in the past added to the confusion by their changing and uncertain use of the names Pinus and Abies. A definite system of naming

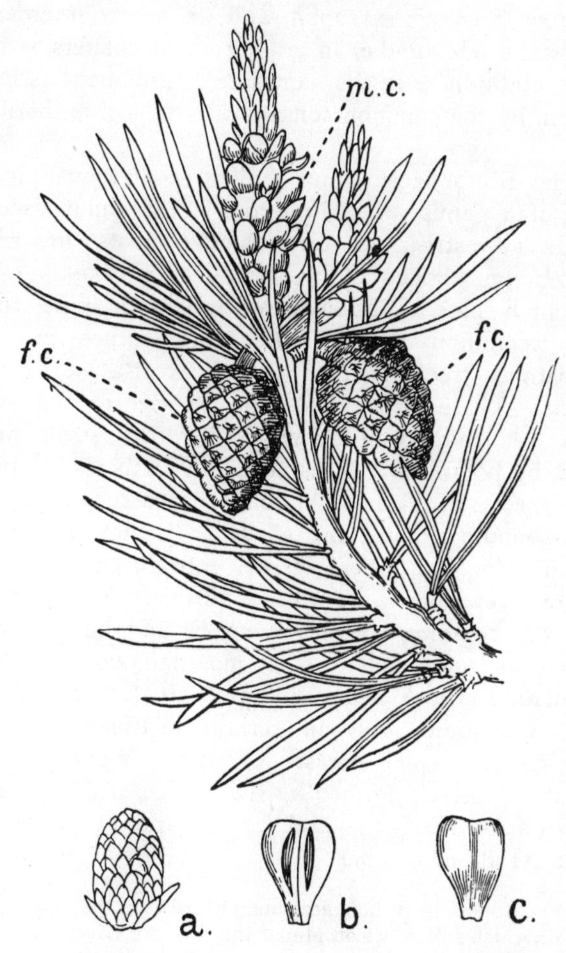

FIG. 31.—A fertile branch of the Scots Fir, Pinus sylvestris, showing the small male cones, *m.c.*, and the larger female cones, *f.c.*, also the foliage needles grouped in pairs. Drawn of two-thirds the natural size, linear.

The lower figures relate to the male cone. a, A ripe male cone, slightly enlarged; b, inner face of one of the scales of the male cone, showing the paired pollen-producing or stamen-like patches — much magnified; c, outer face of the same scale.

has now been agreed upon, and we must, in order to understand one another in talking about conifers, strictly accept and adhere to the names at this moment assigned to them by the common consent of botanical authorities.

The Scots fir is Pinus sylvestris. "Pinus" is the name of a genus of conifers, and includes many species besides sylvestris, our own familar Scots fir, which is often now spoken of by the queer, ill-sounding title of Scotch pine. The Norway spruce or pine, called often "common spruce," also "the spruce fir," and "Christmas tree," is the "Picea excelsa" of correct botany. There are several other species of the genus Picea. A third well-known conifer, the silver fir, is called by botanists "Abies pectinata"; there are many other species of Abies. Although it has such a familiar, sweet-sounding name, the silver fir is not a common tree in England, where it was introduced only three hundred years ago. It will not thrive at Kew Gardens. It is the common forest-making fir of the centre of France and of much of the mountainous country of Southern Europe,[1] but it is rarely to be seen in the Swiss mountains (only in certain relatively low-lying valleys). The pine forests of those mountains are almost exclusively formed by the spruce, with the addition of a few Scots firs and larches, and in some parts of the Arolla fir or pine.

[1] It is, according to botanical authorities, from the wood of the silver fir, which still grows on Mount Ida, that the Greeks, as related by Virgil, constructed the Trojan horse.

"Instar montis equum, divina Palladis arte
Ædificant, sectaque intexunt *abiete* costas!
(A horse of mountain size they build
By art divine of Pallas helped
And weave its ribs with planks of fir).
"Æneid," ii. 15.

The common larch is a fourth common kind of conifer. It is distinguished from other pine trees which flourish in England by shedding its needles so as to leave itself bare in the winter. It is called "Larix Europœa," and is closely related to the cedars. It was introduced into England in 1629.

Man by his migrations and trading journeys has had far more to do with the introduction and spreading of trees, and even of small flowering plants, from one country to another, than is commonly suspected. It appears that of the trees I have already mentioned only the Scots fir is really native to these islands. Even the Christmas tree, the common spruce, was introduced from the Continent by invading man after we had become separated by the sea from the mainland of Europe. The introduction took place, it seems, in very early times, and there is no record of the event. Peat deposits have been studied and their age estimated, and it is found that in those of the age of the neolithic men there are no remains of spruce, but only of Scots firs!

The conifers are remarkable not only for their "cones," but for the needle-like shape which their leaves often present, whence the latter are spoken of simply as "needles." Conifers are also distinguished by the fine aromatic oils which they produce in these needles and in their wood, which serve them as a protection against browsing animals, although to man their perfume is agreeable. In the Tyrol, near Cortina, I remember a little shop in the pine woods where you could buy the odorous essences extracted from the different species of conifers growing around, and each species had its own special perfume. Besides these aromatic oils, the conifers produce peculiar resins, such as colophon, amber, kauri gum, Canada balsam, Dammar varnish, and others, and also various qualities of turpentine, tar, and pitch.

I have mentioned the three commonest conifers which flourish in England, and have pointed out that only one of them—the Pinus sylvestris, or Scots fir—is really indigenous to our islands. It extends all over Europe, except the extreme south and west, and right through Russian Asia. In the Alps, at the height of 3000 to 5000 feet, it is represented by a dwarf recumbent species, the Pinus montana, or P. pumilio. There is another really native conifer in Britain which belongs to a peculiar family, that of the cypresses. This is the common juniper, called by botanists "Juniperus communis," a mere shrub, but still a beautiful little thing, noticeable for the fine perfume of its leaves, which is used for flavouring "gin," and for its peculiar minute and compact berry-like cones. It has a very wide range, flourishing throughout the north temperate region of Europe, Asia, and America. There is another juniper well known in England, namely, the Savin (Juniperus Sabina). This is not a native, but was introduced before 1548. It has powerful medicinal properties.

When we spend our holidays abroad in Switzerland or on the Mediterranean shores we come across many other flourishing, well-established kinds of pines, firs, and cypresses. And we need not leave England in order to make acquaintance with a very large number which have been introduced from abroad into plantations and parks, and grow under favourable circumstances, but cannot be said to have established themselves as naturalized inhabitants. Among those more anciently introduced is the cedar of Lebanon; of later introduction we have the Indian cedar or deodar, and the Weymouth pine, Pinus Strobus, a North American tree. Still later a veritable crowd of American, Himalayan, Japanese, and Chinese pine trees of one kind and another have been introduced by dealers and their rich clients, the owners of park plantations, so that it is now far easier to see in the grounds around great English houses all sorts of pine trees from remote regions of the earth than the British species, or those interesting European kinds which have some kind of community with them, and are, at any rate, objects of interest to the naturalist whose familiar ground is that of Europe. Most people are utterly perplexed by the number of kinds, and do not know one from another.

In order to discuss a little further in detail the commoner kinds of Coniferæ besides those which may be considered as truly British, and have been mentioned above, we must take a glance at the plants related to the natural order Coniferæ, and then at the divisions of that natural order into families and tribes. The Coniferæ are an order of the great class of Gymnosperms—one of two classes into which the flowering plants or Phanerogams are divided, the other being (as explained above) the Angiosperms (palms, grasses, lilies, and all our ordinary trees, shrubs, and flower-bearing herbs). The orders included under "Gymnosperms" are: First, an order, the Pterido-spermia, comprising certain remarkable fossil forms connecting them with ferns; second, the order Cycadeæ, an ancient group, of which only a dozen or so kinds survive to this day; third, the order Gnetaceæ, including Wellwitch's strange African plant and the little European Ephedras, resembling the plants called horse-tails; fourth, the order of the Gingko trees of Japan, called also Salisburiæ, with leaves like those of the maiden's-hair fern. They and one or two others are survivors of an important extinct group (the Gingkoaceæ), which we know by their fossil remains flourished in great numbers before the chalk period. Then we have: fifth, the order Taxaceæ (or yew trees); and, sixth, the order Coniferæ (or cypresses, pines, cedars, and firs). The first four orders, though very interesting, exceptional plants we will leave aside, as they do not come very near to the Coniferæ. The order of yew trees, Taxaceæ, however, does come close to the Coniferæ, and sometimes they are grouped together.

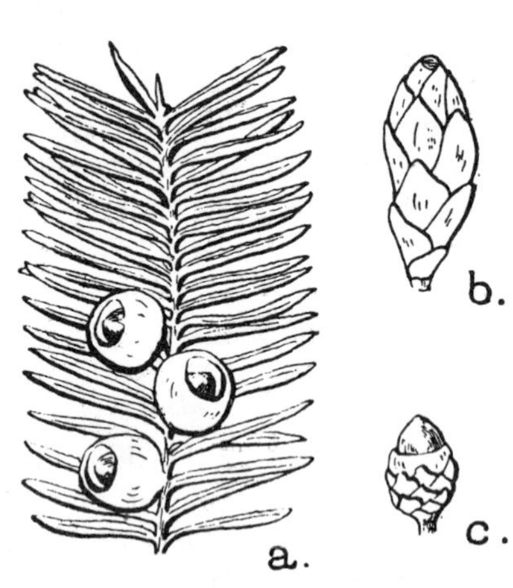

FIG. 32.—The Common Yew, Taxus baccata.

a, Part of a branch (of the natural size) showing the needle-like leaves in two opposite rows, and three fruits. The cup-like growth which is seen surrounding the naked seed is called an "aril." It is of a fine crimson colour, soft, juicy, and sweet-tasting.

b, The young cone-like growth or "flower" of the yew, from the end of which one seed and its cup-like aril will develop. Magnified.

c, The seed surrounded by the incompletely grown aril or cup at a later stage. Less highly magnified than b.

There is one truly native British example of the order Taxaceæ—the common yew tree, called "Taxus baccata" by botanists. Its leaves are "needles," like those of most conifers, but much flattened, and it has the sombre colour and the general aspect of some of the larger conifers. But its ovule-bearing flower, although it appears when young (Fig. 32, b) to be built up by several scale-like leaves like the cone of a conifer, does not continue in that form, and ceases to have any resemblance to a "cone." Only the terminal leaf or scale of the group enlarges and develops an ovule, and around this grows an open cup-like protection of the most delicate crimson colour —soft, sweet, and luscious (Fig. 32, c and a). It is as big as a pea, and is largely eaten by birds and by schoolboys! Yew trees have from time immemorial been planted and cared for in Great Britain, since its wood was formerly greatly valued for making archers' bows. Wild groves of yew trees, once existing, have been largely destroyed. Some of the finest are on the chalk hills of Surrey, where the yew flourishes alongside of the juniper. Very fine yew trees are often found growing, one or two together, in village churchyards, where they have been planted in remote times, just as cypress trees are to-day planted in cemeteries in the South of Europe. Yew trees with trunks from 30 to 50 feet in girth at 12 feet from the ground are known, and it is probable that some are as much as a thousand years old.

Many varieties of the yew tree occur in these islands. A celebrated variety is that in which the branches are all directed upwards rather than horizontally—a frequent form of variation in trees which more usually have spreading, nearly horizontal branches. This variety is called "fastigiate" (the "fastigiate" condition of the common cypress tree is the one usually cultivated, although there are common varieties with spreading branches), and in the case of the fastigiate yew it is accompanied by a variation in the disposition of the needles or leaves. Instead of being carried right and left in a single row on each side of the young branches, as is usual with yews, the needles are set all round the branch in spiral order (as they are in many conifers). This variety was found growing wild in Co. Fermanagh,

Ireland, nearly two hundred years ago, and a couple of trees of it were then cultivated at Florence Court by the Earl of Enniskillen of that date. Thousands of cuttings have been sent from one of these two original trees, which is still vigorous (I saw it some thirty years ago at Florence Court) all over the world. It is known as the "Florence Court yew," or "Irish yew," and is commonly planted in gardens. But all are from cuttings of this one original tree, or cuttings of its cuttings, and all, like their parent, are female berry-bearing trees, for the male and female flowers grow on separate trees in the yew.

The foliage of the yew contains aromatic and other chemical products, which render it poisonous to cattle. It is said not to be poisonous when quite fresh, but only some time after cutting. This, however, needs confirmation. The yew makes an admirably compact and impervious screen when grown as a hedge, and has been largely used in gardens for this purpose. In the sixteenth century it was the custom to clip yew hedges, or small yew trees, into all sorts of strange shapes, birds, beasts, and crowns. The name "topiary" is given to this fanciful work. The popularity of the yew in the gardens of those days is due to the small number of our native evergreen shrubs and trees; they are yew, Scots fir, juniper, holly, privet, ivy, butcher's broom, box (a doubtful native), spurge-laurel, and mistletoe. Up to the end of the seventeenth century only a few evergreens had been introduced from abroad, viz., spruce pine, silver fir, stone pine, pinaster, the cedar of Lebanon, savin, arbor vitæ, evergreen oak, sweet bay, Portugal laurel, laurustine, and arbutus.

I have often wished to have some simple, straightforward information as to conifers, so as to be able to know what differences among them are really recognized by botanists, and what are the correct names of those which one commonly sees. Having gathered that information, I propose to impart it, as far as may be consistent with brevity, to my readers, though I am afraid that to some it will prove a dull business. The order Coniferæ, from which the yew trees (Taxaceæ) are excluded, is divided into four families. These are: (1) the family Abietinæ, which comprises the true pines, and fir trees, and the cedars; (2) the family Araucarianæ, which includes the Monkey puzzle of South America and Australia, and the Dammar tree of New Zealand; (3) the family Taxodinæ, which is best known by the so-called Wellingtonia, or Sequoia, but includes several other genera and species; and (4) the family Cupressinæ, in which the juniper, cypress, and "arbor vitæ," or Thuya, are placed.

The form and size of the frequently needle-like leaves of coniferæ are not of so much importance in indicating the affinities of these plants as one might expect, although their grouping either in tufts or in rows is a matter of significance. In some of them the "needles," or leaves, are long and narrow (Abietinæ); in others they are broad and leaf-like (Araucarianæ); in others they are all or most of them reduced to mere ridges or short scales set quite closely to the leaf-bearing branch (many Cupressinæ and Taxodinæ). It is not possible to give, without going into botanical minutiæ, the items of structure by which the four families of conifers are distinguished from one another. It is best for the nature-lover who is not an adept in botanical details to think of them as grouped each round one well-known species. Thus the Abietinæ are grouped round the spruce pine, the Araucarianæ round the monkey puzzle, the Taxodinæ round the Wellingtonia, and the Cupressinæ round the juniper. In all but the last family the ovule-bearing scales of the female cone are arranged spiral-wise around a central supporting stem ; in Cupressinæ they are few in number, very thick, and opposite to one another so as to form a globular rather than a cone-shaped body. In all but a few Cupressinæ and Araucarianæ the male and female cones are carried on the same tree, sometimes on

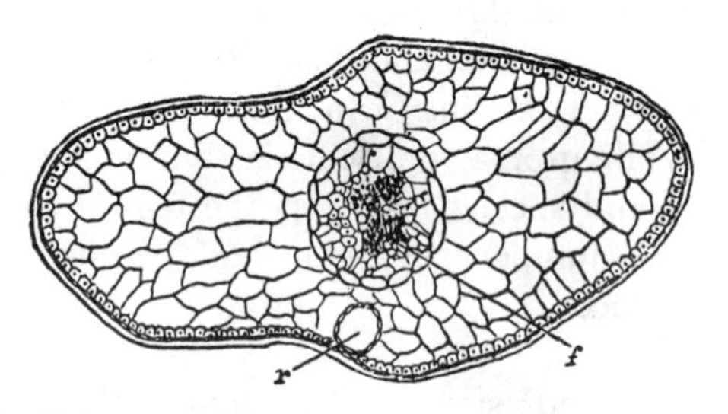

Fig. 33.—A thin slice across one of the foliage needles of the Common Spruce. Highly magnified. *r*, The single resin canal ; *f*, the mid-rib, with a single bundle of fibres and vessels cut across.—(From Veitch.)

separate branches, but usually on the same branch. The male and female cones are always distinct, and the female much the larger and more enduring.

The Abietinæ are divided into three tribes—(*a*) the spruces and silver firs (this group corresponding to the French *Sapins*), (*b*) the larches and cedars, (*c*) the Scots firs (*Pins* of the French). Let us take first the group of spruces and silver firs. The Norwegian spruce is the type of the genus Picea. It is called *Pesse* by the French, *Fichte* by the Germans, and

Picea excelsa by botanists. We may contrast it with the silver fir Abies pectinata (*Sapin des Vosges* of the French, *Silbertanne* of the Germans), which we take as the type of the genus Abies. In many respects the silver fir looks like the spruce. In both the stem is straight, reaching a height of 100 to 150 feet, regularly furnished with tiers of branches from the ground upwards. The leaves are needles, half an inch to an inch long, which stand out from the branchlets, but in the spruce they are quadrangular, green all over, and arise all round the branch, whilst in the

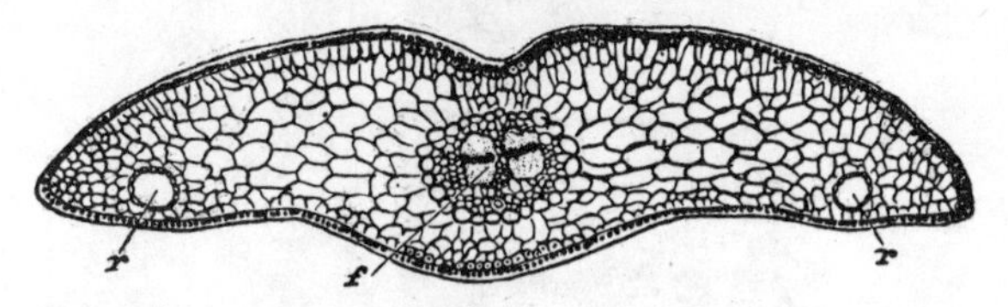

FIG. 34.—A thin slice across one of the foliage needles of the Silver Fir. Highly magnified. It is flatter than the similar slice of the needle of the spruce. *r, r,* The *two* resin canals; *f,* the mid-rib, in which *two* bundles of fibres and vessels can be distinguished.—(From Veitch.)

silver fir they are flat, grooved on the lower surface, which is silver-grey in colour, and they tend to be disposed right and left in two rows. Each needle has a single resin canal in the spruce, but has two in the silver fir, as may be easily seen by cutting the needles across the length with a sharp knife (Figs. 33 and 34). Each scale-like ovule-producing leaf which goes to build up the ripe seed-bearing cone has (as in all conifers theoretically) an outer scale, called a "bract," attached to it which is very short and hidden in the case of the spruce cone, but is longer than the ovuliferous scale, and very obvious in the

FIG. 35.—The upright spine-bearing cone of the Silver Fir, Abies pectinata. The cones vary from this size to one-third as long again. (Copied from Veitch's "Manual of the Coniferæ," by kind permission of Messrs. Veitch.)

silver fir (Fig. 35). It has a triangular re-curved point, which gives the cones of that species a characteristic appearance (Fig. 36). The cones of the silver fir (5 to 6 inches long and 2 inches thick) are set upright on the branches, and when they have shed the seeds the scales fall off rapidly and leave the axis bare,

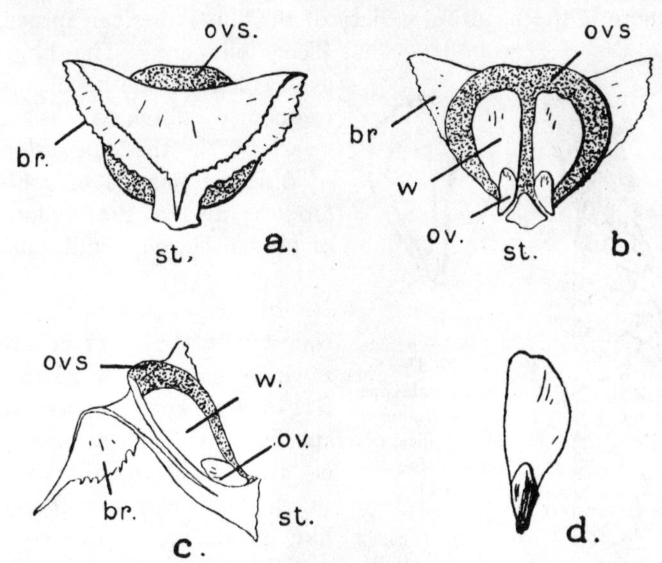

FIG. 36.—a, Structure of the female cone of the Silver Fir. A single cone-scale, OVS, with its reflected and pointed "bract," br, seen from the outer surface; st, stalk.

b, The same seen from the inner surface; letters as before, and in addition OV, one of the pair of naked seeds; W, its wing.

c, The same in section, showing well the reflected or turned-over spine-like end of the bract; letters as before.

d, One of the winged seeds detached.

whilst the cones of the spruce (about an inch shorter) are pendulous (Fig. 37), and their scales remain in position after the seed is shed.

There are many "spruces," other species of the genus Picea, from various parts of Europe, temperate Asia, and North America, which are cultivated in English parks and gardens. Such are the American white and red and black spruces, the Siberian, the Oriental, the Servian, and the tiger's-tail Japanese spruce. Then there is the beautiful variety of the blue American spruce, Picea pungens. The blue-grey colour of the needles is frequently obtained as a "variety" in the cultivation of different species of conifers, as also is the yellow, or golden-leaved, condition.

FIG. 37.
A, The female cone of the Common Spruce, Picea excelsa. Half the natural size (linear measurement). It hangs from its attachment instead of standing up as does the cone of the silver fir.
B, Ripe cone-scale of the Common Spruce, detached and seen from the inner face, so as to show the two winged seeds. Enlarged.

In the genus Abies, associated with the silver fir, are a whole series of American, Siberian, and Japanese species. An interesting one is the Californian Abies bracteata, which has thorn-like processes on the cone 2 inches in length, corresponding to the re-curved spines on the cone of the silver fir. It was introduced into England in 1853, and specimens are growing in Eastnor Park, near Ledbury. The beautiful *pinsapo* of the Spanish Sierra Nevada also belongs to the genus Abies, and may be seen in some English plantations. The Tsuga firs of Japan and North America are related to Abies, but are now placed in a separate genus (Tsuga), as also is the Douglas fir of North America (Pseudotsuga), which has been extensively planted in Great Britain.

The Douglas fir is readily recognized by the decorative trifid outer scales or "bracts" of the rather short cone (Fig. 42). When freshly grown these cones have beautiful purple tints mingled with pale green.

The larches and cedars form the second group or section of the Abietinæ, distinguished by the fact that the needle-like leaves grow in tufts of twenty to forty at the end of short stumpy branchlets or "spurs" (Fig. 38). In the larches, which form the genus Larix, the needles fall off every autumn and leave the tree bare, the annually - renewed feathery foliage contrasting, by its fresh bright green colour, with the darker hues of the persistent needles of other conifers. The common larch (Larix Europœa) is a native of the mountainous regions of Central Europe. The French call it *Mélèze.* There are Himalayan, Japanese, and North American species. The common larch when full-grown is 100 feet and more in height, and has the branches arranged in whorls of diminishing length, so as to give the "Christmas-tree shape" so common among coniferæ. It was introduced into England in the seventeenth century.

FIG. 38.—Cone and foliage (many needles in each tuft) of the Common Larch, Larix Europœa. Of the natural size.

The cedars closely resemble the larches, but have the leaves or needles persistent, and the large cones take two years to ripen, instead of one year, as in all the conifers which I have hitherto mentioned. The cedars form the genus Cedrus, and three species are distinguished, namely: (1) C. Libani, the cedar of Lebanon; (2) C. Atlantica, the North African cedar of the Atlas mountains; and (3) C. deodara, the Himalayan cedar or deodar. They are now considered to be geographical varieties of one species. They differ chiefly in the set of the branches and foliage. The cedar of Lebanon has the trunk forked, and gives rise to large, unequally disposed branches, spreading horizontally; it may have a spread of 100 feet and a height of 70 feet. In this country it is often uprooted by the wind, or its branches are broken by a weight of snow, when it has attained nearly full growth. The deodar cedar is more Christmas-tree-like in shape, the trunk rarely is forked, and it attains, in its native mountains, a height of 250 feet. The Atlas cedar is in many respects intermediate in character between C. Libani and C. deodara. The cedar of Lebanon is undoubtedly the most majestic of the conifers grown in English parks. It was introduced in the year 1665. There are specimens growing in this country of which the trunk has a girth of 25 feet.

The third section of the family Abietinæ is formed by the genus Pinus, of which the Scots fir, or Scotch pine (Pinus sylvestris), is the type. The Abietinæ of this genus are distinguished by their foliage. There are two kinds of leaves—the primitive ones, which are little, scale-like, green up-growths closely scattered on the young branches; and the secondary ones, which are long needles carried as a tuft or fascicle on a very stumpy branchlet. These tufts of needles are persistent (that is to say, are not shed yearly), and differ from those of the larches and cedars in consisting of but few needles in a tuft, the number being characteristic of different species, some having five, others three, others two, and the American Pinus monophylla having only one. The general shape of these trees is not tapering like the spruce with unforked trunk, but they usually shed the lower branches as growth goes on, and present in most cases a trunk carrying an umbrella-like expanse of foliage-bearing branches, or several such expanses. The scales which form the cones in the genus Pinus are (with few exceptions, such as the Weymouth pine) not flat and flexible, but are thickened, swollen, and even knob-like and wooden at the exposed part, which is armed with a weak or a strong prickle (see Figs. 39, 40, and 41). The cones do not ripen until the end of the second or third season; they may be, according to species, erect, pendulous, or horizontal, and vary in size in different species. In some they remain closed on the trees for an indefinite period (even fifteen or twenty years), until opened by the heat of a forest fire or of an exceptionally hot season.

The Scots fir, Pinus sylvestris (Fig. 31), called *Pin de Genève* by the French, has a very wide range. It extends eastward and northward from the Sierra Nevada, in Spain, through Europe and Russian Asia; its northern limit approaches the Arctic circle, its southern limit is formed by the great mountain chains of the Alps, Caucasus, and Altai range of Asia. The beautiful blue-green colour of its needles, the fine red-brown tint of its trunk and branches, and the graceful spread of its foliage high up on a few great, unequally-grown branches springing from its tall, bare trunk, are amongst the most picturesque features of English landscape. In the southern counties "clumps" of a dozen or score of these graceful trees are often to be seen on some isolated hilltop in the moorlands, and are associated with poetic tradition and ancient superstition. In the North of Britain they are more frequent as forest. The Scots fir is the only pine tree really native in our land. It is distinguished from several other species of Pinus by having the leaves or needles in bundles of two, and having relatively small oblong cones (2 to 3 inches long) which are borne near the ends of the branches (Fig. 31). The constituent scales of the cone are only slightly thickened, and the surface knob has no prickle. There are two of the common pine trees of the Mediterranean coast (the Riviera and elsewhere), namely, the Aleppo pine (Pinus halepensis) and the so-called Corsican or Austrian pine (Pinus Laricio), which agree in the above-given points with the Scots fir, and are, in fact, difficult to distinguish from it, except by general shape, mode of growth, and the colour of the leaves and stem. The needles of the Scots fir are 1½ to 3 inches long, those of P. halepensis 2½ to 3½ inches, and those of P. Laricio 4 to 6 inches long. The Pyrenæan or Calabrian pine is closely similar to these.

A very important and abundant pine on the Mediterranean and Biscay coast of France is the Pinaster (Pinus pinaster), often called the "cluster pine," and by the French *Pin des Landes* and *Pin maritime* (Fig. 39). It also has its needles, often 6 inches long, in groups of two. It is usually a smaller tree than the others, but in favourable localities attains a height of 80 feet. Its cones are twice as long as those of the Scots fir, often, as at Bournemouth, 4 and even 5 inches long, and its branches are slender in proportion to the trunk, the bark coarse and fissured, and its foliage (as is that of all the two-leaved set except the Scots fir) of a yellowish (not bluish) green. It has been found invaluable in holding sandy land from shifting and breaking up, and is planted for this purpose along the coast of the Landes and in other parts of the world.

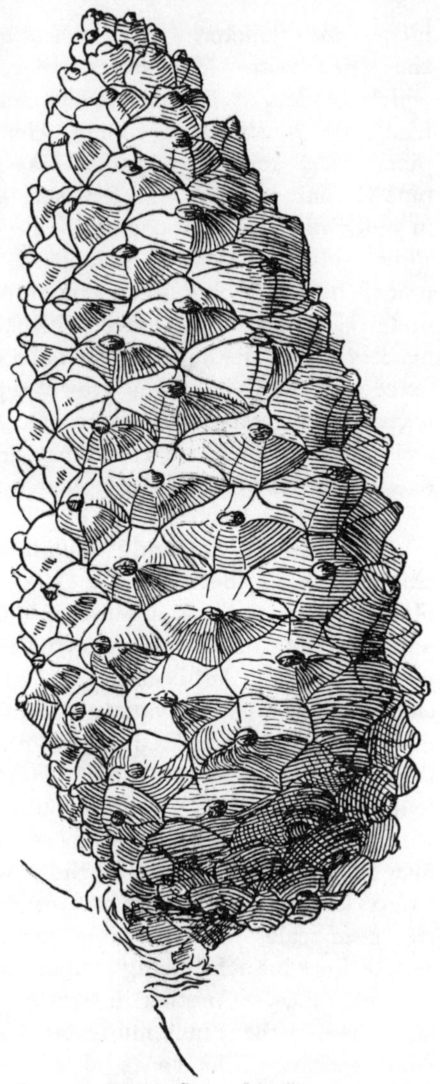

FIG. 39.—Female Cone of the Pinaster, or Maritime Pine (Pinus pinaster). Drawn of the natural size from a Bournemouth specimen.

A still better-known pine, which, like those already mentioned, has its needles in pairs, is the stone pine (Pinus pinea), called by the French *Pin de parasol* and by the Italians *Pino a pinocchi.* This fine tree (usually bigger than the Pinaster) has been largely planted in Italy on account of its picturesque appearance. This is the tree which one sees so often in Turner's landscapes. The needles are 5 to 6 inches long, and the cones are very large and almost spherical, being often 5 inches long and 4 inches in diameter. The cones do not mature until the third year. The scales are very large and solid, which renders it difficult to extract the nut-like seeds, which are roasted and eaten. Hence the name stone-pine. The spreading, parasol-like shape of the stone-pine is characteristic. A few specimens are to be seen in cultivation in this country. In order to distinguish Pinus sylvestris from P. halepensis, laricio, pinaster, and pinea, the deep blue-green colour of the foliage of the first is sufficient, together with the shortness of its needles. To distinguish the others among themselves (except in the case of well-grown typical examples) it is necessary to examine the cones closely, and often when one comes upon these trees they are, on account of the season, devoid of these distinguishing products.

Wide tracts of sandy moorland in the south of England have been in the last century extensively planted with various species of Pinus, and afford the naturalist an interesting opportunity for comparing one with another. At Bournemouth the plantations are chiefly of the Austrian variety of Pinus Laricio,[1] the Scots P. sylvestris, and the Mediterranean Pinaster. The latter is especially luxuriant there. Here and there I have found other species at Bournemouth. A remarkable one with three needles in a group is the Californian Pinus insignis (Fig. 40), known as the Monterey pine. It has a very large cone which is curiously one-sided in growth, the seed-scales on the side facing away from the supporting branch being larger than those on the opposite face. Another interesting species to be met with there is the Pinus muricata, also a Californian sea-coast species. The cones of this species are about 3

[1] A fine specimen is growing near the main entrance of Kew Gardens.

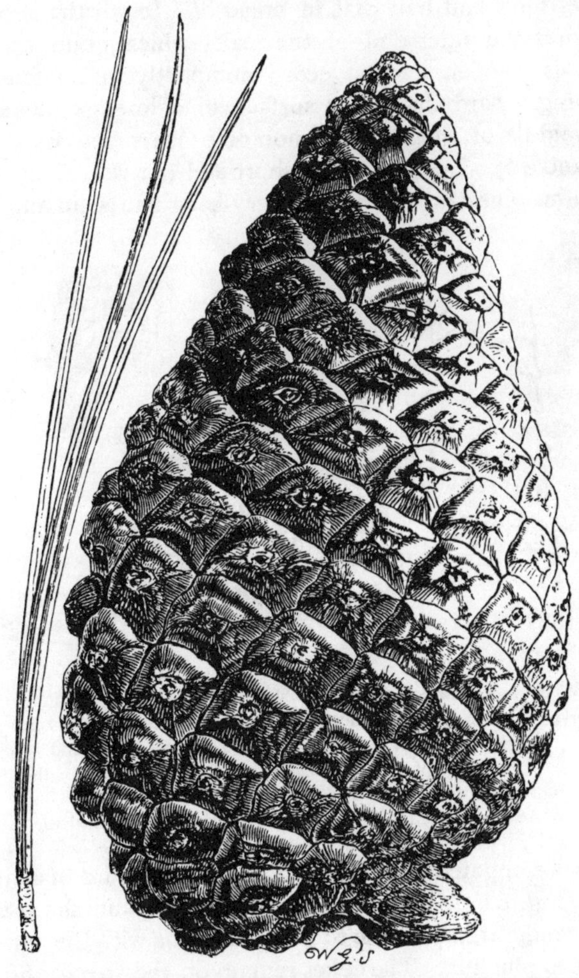

FIG. 40.—Female Cone of the Monterey Pine of California (Pinus insignis). Of the natural size, but somewhat larger specimens are frequent. The characteristic three foliage needles in a bunch, in place of two or five distinctive of some other species of Pinus, are shown in the drawing on the left.—(From Veitch's "Manual of Coniferæ.")

inches long and half that in breadth. In all the species of Pinus the outer end of the scales which build up the cone is swollen and squeezed compactly by its fellows, forming a hard shield-like surface of a lozenge shape, in the middle of which is a knob or process (see Figs. 31, 39, and 40). Usually this is short and not very sharp, but in Pinus muricata the cone is very hard and solid and the

FIG. 41.—Female Cone of Pinus muricata, showing the long sharp spines which stand up from the boss or umbo in the centre of the swollen, woody, lozenge-shaped end or "apex" of each seed-scale. Compare these with the un-armed bosses in the centre of each lozenge building up the surface of the cones drawn in Figs. 31, 39, and 40.

knob is elongated into a spine of nearly one-third of an inch long (Fig. 41). Theses pines are so hard and sharp that they render it impossible to grasp the cone with the hand in order to pluck it. The cones remain on the tree for fifteen years or more, and may be seen in close-set clusters surrounding quite old branches. The cones of Pinus rigida—one of the American pitch-pines—are similarly protected by spines. Pinus rigida is easily distinguished by its having its needles in bundles of three from Pinus

FIG. 42.—Female Cone of the Douglas Fir of North-West America (Oregon and Vancouver), Pseudotsuga Douglasii. Of the natural size.—(From Veitch.)

muricata, which has the more usual arrangement of a pair of needles to each bundle. The Douglas fir is also to be found here and there in the gardens and parks of Bournemouth. Its cones (Fig. 42) are remarkable for their beautiful purple and pale green tints when young, and for the long trifid bract on the outside of each scale, similar to but larger than those on the cone-scales of the silver fir, Abies pectinata (Fig. 35), and not bent backwards as they are.

There are two pine trees of the genus Pinus which one comes across, either in English plantations or on the Continent, and are readily distinguished by having the leaves (needles) in bundles of five. The first of these is the Arolla pine—Pinus Cembra (French, *cembrot*)—a pine tree much like the Scots fir in general appearance, but distinguishable from it, not only by the tufts of five needles in a bunch instead of two, but also by the erect cones which are nearly as broad as long (3 in. by 2 in.). It is essentially a Siberian tree, and grows in Europe only on the Carpathian Mountains and the Alps. I have seen it in the neighbourhood of the Rhone Valley in Switzerland, but it is yearly becoming rarer owing to its destruction at the great heights (4000 to 6000 feet), where it formerly flourished, by the herdsmen in order to extend the pasturage for their milk industry. The other pine with five leaves in a tuft, which one may often see, is the Weymouth pine—Pinus Strobus. It is a native of the New England States and Canada, where it is known as the white pine, and is greatly valued as a timber tree. It was introduced and planted in England by Lord Weymouth at the beginning of the eighteenth century, and is a very handsome tree, growing to 120 feet in height, with a bluish-green colour of the foliage like that of the Scots fir. The needles are 3 to 4 inches long, and the cones pendulous, 5 to 6 inches long and blunt. Another pine of the five-leaved group is to be seen in gardens in the South of Europe (for instance at Baveno on the Lago Maggiore), where it is introduced from Mexico. This is the Pinus Montezumæ, which has extraordinarily long tufts of needles of a blue-green colour, each needle from 7 to 10 inches long, arranged as radiating or fan-like growths of great beauty and striking appearance. The Bohtan pine of the Himalayas (Pinus excelsa—not to be confused with Picea excelsa, the spruce) is also a five-leaved species. Several specimens of it are flourishing in Kew Gardens.

A few lines must be given to the Araucarianæ, Taxodinæ, and Cupressinæ. The Araucarianæ include, besides the Chilian monkey puzzle, an Australian species, and the New Zealand Dammar pine Agathis, which produces the amber-like Kauri gum. The leaves of the monkey puzzle are like the scales of a spruce cone in shape, and the ordinary branches are like elongated green spruce-cones, whilst the seed-cones have needle-like scales. The next family, the Taxodinæ, are in many respects intermediate in character, between the Abietinæ (true pines, cedars, and firs) and the Cupressinæ (cypresses and junipers). They have very small, lance-shaped leaves, closely packed, so as to overlap one another—as in the celebrated Wellingtonia or American Big-tree—and small cones, with hard, knob-like scales, resembling those of the most woody-coned Pinus, but few in number. The American Big-tree (native on the western slopes of the Californian Sierra Nevada) is named "Sequoia gigantea" by the botanists. It was introduced into England about sixty years ago. The Red-wood, of the Pacific coast of the United States, is another species of Sequoia (S. sempervirens), and it appears that a specimen of it has been measured as reaching 340 feet in height;

whilst no living specimen of the S. gigantea has been definitely measured of more than 325 feet in height. There are several other large exotic, pine-like trees, which are placed in the Taxodinæ. The extraordinary and interesting tree called the Japanese umbrella pine (Sciadopitys verticillata) is associated with the Sequoias by some botanists; but it is in important respects unlike any other conifer. It has a very peculiar foliage, namely, rod-like leaflets, twenty to thirty in number, arranged in circlets or whorls like the spokes or ribs of an umbrella. The curious thing is that these are not "leaves," but, according to botanists, are leaf-like shoots or branchlets! It may be seen growing in Kew Gardens, where it was introduced thirty years ago.

The last family of the Coniferæ is the Cupressinæ, so named after the great and beautiful cypress tree, which is said to have given its name to the island of Cyprus, which in turn gives its name to cupreous metal, or copper. The cypress tree similarly gives its name to "coffers" and "coffins" made of its wood, as the Buxus or box-tree has given its name to a "box." The cypress is the Gopher tree of the Hebrews. The family includes many species of junipers (Juniperus) and the American and Japanese Arbor vitæ (Thuya) and its allies. In the common cypress (Cupressus sempervirens) the leaves are singular, small, scale-like growths, which are flattened on to the delicate branchlets which bear them. In other trees of the family both such leaves and also upstanding lancet-like leaves are present. The main character is the small size and globular shape of the cones and the very few swollen scales, more like solid wedges adherent to one another, which build them up. These wedge-like scales are not arranged in whorls, but are opposite to one another on the short axis or stem of the cone. The common juniper

(Juniperus communis), the *génévrier* of the French, grows abundantly on the chalk downs of the South of England, where it appears as a small bush, not exceeding 5 feet in height, but in favourable conditions reaches a height of 20 feet. The cones of the juniper are numerous, and each consists of only three ovuliferous scales, and is only one-fifth of an inch in diameter when ripe, and of a blackish violet colour.

At the close of this compressed survey of the order Coniferæ, let me put the chief forms and groups at which we have looked in a tabular form, thus:

Order CONIFERÆ:

Family 1.—ABIETINÆ.

Section A.—Sapineæ (Spruces and Silver Firs).

Genus 1.—Picea. 2. Tsuga. 3. Pseudotsuga. 4. Abies.

Section B.—Lariceæ (Larches and Cedars).

Genus 1.—Larix. 2. Cedrus.

Section C.—Pineæ.

Genus unic.—Pinus.

Family 2.—ARAUCARIANÆ.

Genus 1.—Araucaria. 2. Agathis. 2. Cunninghamia.

Family 3.—TAXODINÆ.

Genus 1.—Sequoia. 2 Taxodium. 3. Sciadopitys, etc.

Family 4.—CUPRESSINÆ.

Genus 1—Cupressus. 2. Thuya. 3. Juniperus, etc.

CHAPTER XXXIII

THE LYMPH AND THE LYMPHATIC SYSTEM

MOST people do not know even of the existence in their own bodies of a fluid called "the lymph," and of a system of vessels and spaces containing it which ramify like the blood-vessels into every part of the body. This arises from the fact that the lymph is translucent and colourless. You can see the finest blood-vessels when the body of a dead rat, sheep, or man is opened, because they are filled with the beautiful red blood, and appear as a rich, coloured network. But the lymph and the lymph-vessels escape notice, and, indeed, are invisible except the largest, because they are colourless. They remained unknown to anatomists long after arteries and veins, and the fine networks of hair-like vessels or capillaries connecting them, were thoroughly well studied. It is, when one thinks of it, a very noteworthy fact, tending to convince us of the readiness with which we may (in the absence of careful examination and attention) overlook the most weighty things, that here is a great system of vessels and spaces in the human body and in that of other animals, carrying on most important operations in our daily life, and yet most of us have never seen any evidence of its existence, and never hold it in our mind's eye as part of the great mechanism of the animal body.

The lymph is a clear, colourless fluid, with "corpuscles"—minute nucleated cells or particles of protoplasm — floating in it. The liquid part is closely similar in its properties and chemical constitution to the liquid part of the blood. It, indeed, consists largely of the liquid part of the blood which exudes from the finest hair-like blood-vessels or capillaries as they traverse the various tissues, and it is the chief business of the "lymphatics" or lymph-holding vessels to return this exuded liquid to the blood system, which they do by joining—like the rivulets of a river system—to form two large trunks which open into the great blood-holding veins at the region where they approach the heart. The total amount of lymph in the lymphatic system is difficult to estimate, but it is larger in quantity than the blood in the entire blood-vascular system. A large number of the delicate vessels of the lymphatic system take their origin just below the lining layer of the intestine, and ramify through the transparent membrane, which holds the coils of intestine together, and is called the mesentery. The fatty or oily materials of food pass through the lining "cells" of the intestinal wall into these "lacteal" or milky lymphatics, and consequently in an animal killed and examined after a meal, the fluid in them has a milky appearance, and renders this kind of "lymphatics" visible.

They were for this reason the first to be detected, and were known even in ancient times to anatomists. The milky fluid in them was called "the chyle." Its milky appearance is due to the same cause as the white opaque appearance of milk, namely, to the presence of an immense number of excessively small particles of oil (fat) and a certain proportion of larger globules of the same nature. It was thus not difficult for the old anatomists to trace the fine branches of the lacteals uniting branch to branch, and at last forming a large trunk—called the thoracic duct—about a quarter of an inch thick, which runs up the inner face of the backbone to the neck, where it joins the great left subclavian vein, and pours its contents into the blood-stream which is there nearing the heart. A small trunk formed by the union of lymphatic vessels from the right side of the head and neck and the right upper limb opens into the right subclavian vein. It took some time to discover this smaller trunk, since it is not brought to view by milky contents. Gradually it was made out that there are innumerable transparent branches opening into the thoracic duct from the whole of the body, besides the milky-looking lacteals: branches which bring "limpid" clear fluid, or "lymph," from all the viscera, from the muscles, and from the deeper layers of the skin in every region of the body, even from the toes, fingers, and tongue tip. In fact, wherever the blood-vessels take blood there are also vessels of the lymphatic system

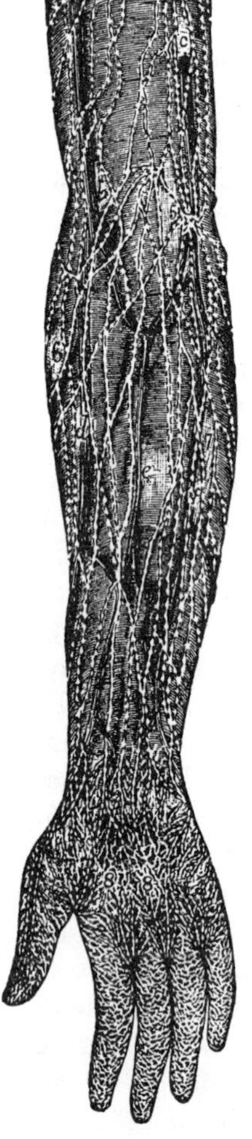

FIG. 43. — The fore-arm of man, with the skin removed so as to show the large superficial lymphatic vessels resting on the muscles. They are represented as white knotted cords. On the palm of the hand (8, 8) and on the fingers a closer network of these vessels is represented, but the smaller lymphatic capillaries and spaces are not shown.

bringing back to the heart the liquid exudation which escapes into the tissues from the finest blood-vessels (Fig. 43).

Whilst we distinguish in an animal body various "tissues" which have special properties and activities, and can be dissected out and delimited—as we could dissect and distinguish the "tissues" (flannel, silk, leather, whalebone, wadding, gold-thread, etc.) making up an elaborate padded, stiffened brocaded, lined, and decorated costume—we find that, unlike what is usual in a man-made costume, all the parts of an animal body (viscera, and their lobes and sub-divisions, the blood-vessels, nerves, muscles, bones, etc.), are covered and separated from one another, and, at the same time, held together by a ubiquitous soft, spongy tissue, consisting of delicate threads and bands, enclosing spaces—some excessively minute and narrow, others larger—in which is a liquid. This is the great packing tissue of the body, and is called "the connective tissue." Its threads and bands have delicate, usually flat nucleated corpuscles (so-called "cells") of transparent protoplasm resting upon them and bathed by the liquid in the fine spaces. The threads and bands are, indeed, the product of the protoplasmic cells, built or "spun" by them, laid down by them as a snail leaves a slimy smear behind it as it crawls. It is not difficult to cut out transparent pieces of this "connective tissue" from a recently killed animal and to examine it with a very high power of the microscope. You may then see the living protoplasmic corpuscles slowly "streaming" and changing shape, and sometimes dividing (one into two) so as to form new corpuscles.

I made my first acquaintance with them when I was

a student at Vienna with the great microscopist Stricker. We used the glass-clear connective tissue which forms the "cornea" of the eye, cut from a freshly killed frog. In those days the part taken by these cells in inflammation was being discovered, the name "phagocyte" had not been invented, the part played by them and by bacteria in disease and the suppuration of wounds was unknown, and I had the privilege of introducing Lister's earliest researches on aseptic surgery and on the coagulation of the blood to the notice of my friend and teacher.

This ubiquitous "connective tissue" underlying the skin, pushing its way into and around every part of every structure in the body, is the "source"—the reservoir, as it were—from which the lymph stream and the finest lymphatic vessels take their origin. The question may very naturally be asked, "How is it that the lymph flows along the channels provided by the transparent lymph vessels and is poured through 'the thoracic duct' into the great vein near the heart?" If we inject a suitable coloured fluid by means of a needle-pointed syringe into any mass of connective tissue, we can see the fluid pass into the numerous lymph vessels previously invisible, and if we inject into them a weak solution of silver nitrate we can, subsequently by aid of the microscope, make out the structure of the walls of the lymphatics and the lining pavement cells which become stained of a brown colour by the silver when exposed to light. But there is no muscular envelope, nothing like "a lymph-heart" in mammals, to drive the lymph along. There are valves or flexible flaps in the walls of the lymph-vessels, as there are in the veins, and the lymph is driven to the heart by the intermittent pressure upon these valved tubes, caused by the movements of the muscles and of the body generally. The valves, like those of the veins, prevent the flow of the lymph backwards, but allow it to pass forward towards the heart. This is shown by the examination of a narcotized mammal (killed immediately after the examination has been made). A glass tube is placed in the thoracic duct, and about a dozen drops of lymph (which would have been delivered into the great vein) pass from it in a minute. If, however, the animal's legs are moved, as though in running, or if "massage" is applied to the limbs—the pressure being directed from the extremities towards the heart—then a greatly increased flow of lymph is observed, as much as sixty drops in a minute! This is the chief explanation of the value to our health of exercise, and also of the importance of "massage" as a treatment in disease. Either exercise or massage entirely revolutionizes the rate of flow of the lymph, quickening it so greatly that the physiological effect on the general chemical processes going on in the body cannot fail to be most important.

Curiously enough, whilst mammals have to depend entirely on pressure and exercise for anything but the slowest flow of the lymph, the cold-blooded vertebrates, fish, amphibia and reptiles (and even some birds), have remarkable, rhythmically contracting, muscular sacs, which pump the lymph from large lymph-vessels into large veins, and are called "lymphatic hearts." The eel and other fish have them in the tail, but they are best seen in the common frog. There is an anterior pair, one under each shoulder-blade, and another pair, one on each hip. Each opens at one end into a large "collecting" lymph-vessel, and at the other end into a large vein. They "beat" like a heart, but do not keep time with one another. Their muscular walls are formed by what is called "striated" muscular tissue (as are those of the blood-heart), and they are under the control of branches of the spinal nerves. The movement of the hinder pair in a frog can be seen through the skin.

In man and all vertebrate animals the intestines, stomach and liver, heart and lungs (or swim-bladder) lie loose, except for a fibrous band of attachment, in a great cavity (often divided into two or more chambers), which they fit fairly closely. The small space between them and the walls of the cavity is occupied by a liquid. This is lymph, and the great cavity is a lymph-space. When this cavity is in its primitive form it is called the body cavity, or "cœlom." In man and mammals it is divided into four chief chambers—the peritoneal cavity (in which the stomach, intestine, and liver are loosely attached and have a certain mobility), the right pleural and left pleural cavity (one for each lung), and the pericardial cavity (for the heart). These great chambers are part of the lymph-system, and so is the lymph-holding space around and within the brain and spinal cord, and so are the great spaces beneath the frog's skin.

If we look at the structure of an earth-worm or of one of the graceful marine worms (Nereis or Arenicola), we gain a good deal of light as to the nature of the lymphatic system of Vertebrates. Suppose you have killed a large earth-worm with chloroform! Then pin it out on a cork plate, and open it by a cut along the back with a fine pair of scissors. The point of your scissors passes through the muscular body-wall of the worm into a great chamber filled with a clear liquid. This chamber is the "cœlom," and is the same structure as the pleural and peritoneal chambers of the Vertebrate. But it holds (proportionately) more liquid. The liquid is "lymph," like that of the Vertebrate, and has numerous protoplasmic cells floating in it. There is comparatively little connective tissue in the earth-worm. The cœlom is free and unblocked—the great viscera lie in it. There are some delicate, transparent bands of connective tissue, but not much nor bulky. The wall of the cœlom itself is lined with connective tissue, and if that tissue grew greatly in bulk, and bound all the organs and muscles together, it would reduce the large cavity, filling it up with spongy tissue in the small interstices of which there would be lymph. And so we should get a lymph system resembling that of Vertebrates, instead of one large chamber.

But what about the opening of the lymphatics into the blood-vessels? This is one of the interesting differences between the earth-worm and the Vertebrate. The earthworm and many marine worms have a beautiful system of vessels, containing a bright red blood, and forming true capillaries, connecting arteries and veins. The heart is a long, rhythmically beating tube, extending along the whole length of the animal just above the intestine. There is no opening into it of the lymph-cavity. It is purely a respiratory blood-system, pumping its fluid, coloured red by oxygen-seizing hæmoglobin into every part of the body. It passes along the fine capillaries of the skin, where it seizes oxygen from the outside air or water and carries it to all the tissues. The fact is that the red respiratory element of the blood which we call the "hæma" or hæmal portion (the Greek word for red blood is *αἷμα*) is here kept separate from the nourishing and elaborating element, the lymph or lymphatic portion. So that we should, to be explicit, describe the blood of a vertebrate as "hæmolymph," a conjunction of hæma and lymph, which in the more primitive earth-worm and sea-worm have neve effected a junction ! In some closely allied marine worms, however, a junction of these two is effected in another way. We know that in the Vertebrates the red blood corpuscles are formed by detached bits of the same tissue, which becomes converted into capillaries, the finest blood-vessels. Now in several marine Chætopods or bristle-footed worms (Glycera, Capitella, etc.) the tissue which should form the blood-vascular system and its red liquid blood, changes its mode of growth ; it never forms blood-vessels at all, but divides into free red (hæmoglobinous) cells or red blood corpuscles, which float in the lymph of the cœlom. There is no blood-vascular system produced in these worms, but the "cells" of the tissue which would in other worms form blood-vessels break up into red corpuscles, which, mixing with the lymph, bring it into the condition of "hæmolymph," identical with the blood of Vertebrates !

In the molluscs, snails, whelks, oysters, clams, and cuttle-fishes there is a further variation. The same two fluids and two systems of spaces are present as in the earth-worm, but the cœlomic space and fluid have been nearly blocked up and obliterated by the swelling-up and great size of the proper hæmal vessels. Only in rare cases is the blood of molluscs coloured red by hæmoglobin, usually it is of a pale blue colour. There is still left a pericardial cœlom, a space around the heart, and from this some fine lymph-holding vessels ramify amongst the tissues, but the chief spaces in the body are dilated parts of the true hæmal system. In Insects and Crustacea (say cockroach and lobster) this process is carried still further. The great cœlom, so well developed in the Chætopod worms, and the Sea-urchins and

Star-fishes, and retaining quite a large development also in the Vertebrates, is nowhere to be found. The swollen blood-vessels have squeezed it out of existence, except for certain sack-like remnants which enclose separately the ovaries, and the testes, and the kidneys, and have each its opening to the exterior conveying the products of those important organs to the outer world. Thus we gain a brief insight into the true history of the lymphatic system and its vicissitudes in the lower animals and in man.

CHAPTER XXXIV

THE BLOOD AND ITS CIRCULATION

RED, crimson, scarlet, hot, the river of life, the carrier of all that is good and all that is bad by its myriad streams through our bodies; the rarest, most precious, most gorgeous of fluids; the daughter of the salt ocean, finer and more worshipful even than the waters of the great mother, the sea; the badge of horror and of accursed cruelty, yet also the emblem of nobility, of generosity, of all that is near and dear, of all that is splendid and beautiful; the blush of modesty and the flag of rage; the giver of coral lips and glowing cheeks to youth and health, and no less of the ruddy nose which women hide with powder and men bravely bear without concealment! Such is the blood, and it is no wonder that the mere sight of it has always had an overpowering fascination for mankind.

The wild people of the Solomon Islands, when they see a drop of blood flowing from an accidental scratch of hand or foot, say, "I must go home; some danger is at hand; the blood has come to tell me!" Sorcerers and witches of all times have endeavoured to procure a few drops of the blood of their intended victims in order to "work spells" upon the precious fluid, and so, according to the theory of "contagious magic," upon the person from which it came. In Italy to-day, as in this country a few hundred years ago, when some one's nose bleeds, a Latin hymn to the blood (beautiful in its conception) begging it to stay its flow, as it did when the soldier's spear pierced the side of the crucified Christ, is sung. In a village in the hills near Naples I was taken with an attack of nose-bleeding, and bathed my head with cold water from a pretty fountain which supplied the people with its pure stream. The women brought handsome old brass basins and embroidered cloths of the most delicate linen for my use. I heard a strange chanting behind my back as I stooped over the water, and when the bleeding had ceased I found that an old man of the village had placed two straws in the form of the cross on my shoulders, and was reciting the ancient Latin hymn to my overflowing blood! I obtained afterwards from a friend the words of the same hymn as used in long-ago days in English villages.

One primitive race if not others, namely, the Australians, take a very prosaic and business-like view of the blood. They use it as an adhesive—a sort of liquid paste or gum, always ready to hand! In order to fasten feathers or other decoration to a pole, the Australian "black fellow," without wincing or hesitation, and as a matter of course, makes a cut (with a sharp piece of stone or glass) in his own arm, and uses the convenient blood. It also serves them as paint, as it has served many a chieftain of European race for signing his name, and many a prisoner for writing in the absence of ink.

There is for some people a fascination in the sight of blood which must not be mistaken for cruelty, although it is accompanied by dangerous and undesirable emotion. Just as other emotion-producing experiences—such as the sight or hearing of torture, of hairbreadth escapes, and of ghosts—produces uncontrollable repulsion and horror in some people, and to others (or even to the same people when in another state of health or mental balance) actually gives a pleasurable sensation (exquisite shudderings, as the French say), so does the sight of blood or even the mere hearing of the word "blood" act differently on different people. Every one who has witnessed a Spanish bull-fight knows that it is not any desire for, or enjoyment of, the sight of pain which excites the crowded mass of spectators. There is no "cruelty," in the proper sense, in their state of mind, no pleasure in witnessing pain—a thing which, terrible as it is to think of, yet does exist naturally in mankind, and has to be, and is, repressed and absolutely got rid of in the course of the humanizing education of civilized mankind. The spectators of the Spanish bull-fight are primarily under the spell or fascination of the sight of blood, and in a less degree they are attracted by the wonderful exhibition of skill and strength on the part of the matador and his troop. The crowd excitedly acclaims the first drops of blood which the splendid bull is made to shed. They buy, after he has been killed, the paper-winged darts smeared with his blood. The colour, the mystery, and the magnificence of blood produces in them a violent emotion. It is to them a delight, but only a single step separates their delight from pain and actual physical distress. The most absolutely nauseating smells are very nearly identical with delightful perfumes, and we all know how readily a taste may be acquired converting the former into the latter—as in the case of the (to most people) foul-smelling East Indian fruit, the durian, and of rotten cheese and "high" game. We also know that a sudden revulsion of "feeling" may occur in regard to hitherto approved smells and flavours, so that headache, vomiting, and even fainting may be produced by a smell or flavour which was previously found a favourite beyond all others.

So it is with this great and mysterious thing—the blood. The sight of it nearly always produces emotion and excitement, but if these emotions are not accompanied by an unreasoning joy and delight, they may result in equally unreasoning and uncontrollable disgust, horror, and often a sudden and unaccountable collapse. Some time ago in a popular lecture on the colouring matter of the blood I had no sooner said the word "blood" than a gentleman in the front row fainted and had to be carried out. Men are more susceptible to this curious effect of the sight or thought of blood than women. Often they do not know that they are so, and are as astonished and perplexed by the sudden fainting as are onlookers and as are, for the matter of that, physiologists and psychologists. It is a common experience of medical men who vaccinate adults, when there is a scare about small-pox, that at the sight of a tiny drop of blood caused by scratching the arm with a lancet, men frequently faint, whilst women rarely do so. Great, burly, red-coated soldiers, and also athletic schoolboys, have been especially noted as fainting when vaccinated. Maid-servants rarely faint under this absurdly trivial ordeal, whilst the butler and the valet much more frequently do so. Here is, indeed, a curious and unexpected difference between men and women which I commend to the consideration of those who are discussing the desirability of admitting women to the parliamentary franchise. It is an unexplained instance of the influence of the mind on the body, and until it is better understood, one must not conclude that the difference is a proof of superior fitness for participation in political affairs.

I trust that none of my readers may suddenly faint on reading this page, but should be glad to hear of any experience of the kind. It is readily understood when the profound impression produced by the colour of man's blood is considered, that the great inquirer Aristotle and a good many uninquiring people of the present day should overlook the fact that the lower animals have blood. The insects, crustaceans, mussels, clams, snails, and cuttle-fish, and many worms have true blood and a heart and blood-vessels, but in most of them the blood is colourless, or of a very pale blue tint. Hence, like the lymph described in the preceding chapter, it escapes attention, and Aristotle called them all "bloodless animals." The fact is, however, that not only do they possess colourless or pale blue blood, but that the bristle-footed worms (earth-worms and river-worms and marine Annelids) and even the leeches possess bright red blood contained in a complete branching network of blood-vessels, whilst here and there among the otherwise colourless-blooded molluscs and crustaceans and insects we find isolated instances of the possession of red blood. Thus the flat-coiled pond-snail, Planorbis, has bright red blood, so have one or two bivalve clams, so, too, has an insect larva (known to boys as a bloodworm) that of the midge (Chironomus), so, too, have some small fresh-water shrimps, and also a single species of star-fish and one kind of sea cucumber!

I explained in the previous chapter that the blood of the vertebrates may well be called hæmolymph, since in them the colourless, slightly opalescent fluid called "lymph" is continually poured through certain openings into the red blood, and mixed with it. In the earthworm and other lower animals the red-coloured blood, or its equivalent—the "hæma," as distinguished from the "lymph"—is held in a closed system of vessels, and does not receive any of the lymph. When examined with the microscope, the blood, or hæmolymph, of man is found to consist of an albuminous, slightly sticky liquid, in which float an immense number of "corpuscles" —minute bodies, some rounded, some irregular, some bun-like, and some spherical. The most abundant of these are the "red corpuscles," of the shape of buns, slightly depressed on each surface. Three thousand two hundred of them could be placed lying flat side by side along the space of a measured inch. They appear pale greenish-yellow in colour under the microscope, but in quantity, lying one over the other, they allow only red and some blue light to pass through them, and so have a fine red colour. They consist of a small quantity of albuminous matter and water, and of a large proportion of a red-coloured, crystallizable, chemical substance dissolved in them, called hæmoglobin, or blood-red. It is this hæmoglobin which performs one of the most important duties of the blood, since it combines with the oxygen of the inspired air when the corpuscles are flowing through the fine vessels of the lungs, and carries it to the tissues in every part of the body, which greedily take the oxygen from the red corpuscles.

The red corpuscles of man's blood and that of the hairy suckling animals—the mammals—are not nucleated cells, but are regularly formed and renewed as they daily wear out, as fragments of larger mother-cells, which break up into these corpuscles, in the marrow of the bones, and some other situations where they are found. In all other vertebrates the red blood corpuscles have a kernel, or dense nucleus, and are complete "cells," usually oval, smooth and flattened in shape—a curious difference not easily accounted for. There are in a pint of the blood of an average man about two billions of these red corpuscles, and the amount of blood in the body is about one-twentieth of the total weight of the body—say, in a man weighing 160 lb., about 8 lb. or pints of blood. The clear, colourless lymph existing in all the lymph spaces of the body is probably about twelve pints. In many animals the red corpuscles are much less numerous than in man; for instance, a drop of human blood contains a thousand times as many red corpuscles as does an equal-sized drop of frog's blood. It is true that the frog's red corpuscles are a good deal bigger than those of man, but the result is that the human blood is some hundreds of times richer in hæmoglobin than the frog's, and has a proportionately greater power of carrying oxygen from the lungs to the tissues, and keeping up the slow, burning process, or oxidation, upon which the activity of the body, as well as its warmth, depend. The body depends upon its supply of oxygen as a steam-engine depends upon the oxygen of the air, which keeps its coal-fire burning.

The pace of the blood-stream which is produced by the force-pump action of the contractions or beats of the heart is tremendous. It courses along at the rate of ten inches in a second in the big arteries and veins, and it has been carefully ascertained by experiment that a heartful of blood (which in a big man is about half a pint for each half or "side" of the heart)—or let us speak of a single corpuscle—is driven out of the heart through the great artery or aorta to the most remote parts of the body, and is back again at the heart, after running through endless branches of arteries, smallest capillaries, and thence into fine veins, bigger veins, and the biggest vein, in twenty to thirty seconds, the time occupied by twenty-five to thirty heart-beats. The walls of the arteries are firm, though elastic, and it is no wonder, with this tremendous pressure and pace on the liquid within, that when an artery is cut the blood spurts out to a distance of several feet.

The colourless liquid of the blood contains, besides the red corpuscles floating in it, others brought to it in the lymph and derived from various connective-tissue spaces and special nodules or "glands." They are outnumbered by the red corpuscles in the proportion of five hundred to one. They are colourless, and bigger than the red corpuscles. Most of them continually change their shape, and consist of active, moving protoplasm. These are the "phagocytes," which, besides acting chemically upon the constituents of the blood-liquid, take into their substance (as does the amœba or proteus-animalcule) and digest and destroy all foreign or dead particles, and the bacteria which may find their way into it. They pass out, forcing their way through the excessively thin walls of the finest capillaries—blood-vessels not wide enough to admit two of them side by side—and enter, to the number of thousands, the tissues which have been wounded or poisoned by bacteria, to carry on their all-important protective "scavenger" or "police-constable" work.

Inflammation is the slowing of the blood-stream by dilatation of the vessels at an injured spot, in order to allow the phagocytes to make their way out of the blood-stream into the tissues, and so get to close quarters with the enemy. There are other excessively minute dust-like particles called "platelets," which are sometimes very abundant in the liquid of the blood. Besides the duties of oxygen-carrying and scavengering the blood has other great and vitally important business. It has to distribute nutriment, to pick up waste oxidized chemical products and get rid of them, and to distribute and equalize the heat which it carries around the body like a perfect hot-water warming installation.

CHAPTER XXXV

FISH AND FAST DAYS

MOST people are familiar with the fact that fasting in the Christian Church has from early times been of two degrees—one in which no flesh of beast or bird or fish, not even eggs, not even milk, may be consumed, and a less severe degree in which the eating of fish is allowed. It is not at first sight clear why the eating of fish—and even of birds such as the Barnacle goose and the Sooty duck, supposed to be produced from fish—has been permitted by the Christian Church, since the flesh of fish is highly nourishing and an excellent substitute for the meat of beasts and birds, and a man fed upon it is far from suffering the effects of true "fasting." Many races and out-of-the-way people live entirely upon vegetables and a little fish, and do very well on that diet.

It has been proved by some learned inquirers that there was a special significance about the permission by the early Christians of a fish diet during so-called "fasting." Real and complete fasting, abstention from all food, for a day or even a week, was and still is practised by some Eastern peoples as a religious exercise. It is a matter of fact that an ecstatic condition of mind is favoured by complete fasting, and conditions favourable to illusions of various kinds are so produced. But the later Christians seem to have regarded the partial fasting during Lent and on certain days of the week as a sort of protest against gluttony and excess, and there is no objection to it among Protestant Churches excepting that it must not be claimed as a merit or the equivalent of "good works."

That fish were, even in the most ancient times, allowed to be eaten on fast days is curious. It is suggested by some students of this subject that the custom came from Syria, and had to do with certain pagan ceremonials and the worship of the fish-god Dagon. It is supposed that some of these early Christians managed, under the guise of a fast of the Church, to maintain an ancient pagan custom and religious rite connected with the Syrian fish-god. The Jews also eat fish on Friday evening—though in both cases the origin of the "fish-eating" was lost sight of in the early centuries of the Christian era. On the other hand, it appears that the worshippers of the fish-god (at any rate, at a remote period) were forbidden to eat fish as being sacred; hence it seems possible that the permission of a fish diet to Christians during days of fasting was given as a means of encouraging those who retained pagan superstitions to ignore and forget them. The supposition that the eating of fish on certain days is a survival of a ceremonial observance connected with fish-worship is the more probable explanation of the custom.

The worship of fish or of a fish-god is one of the outcomes of the old Nature-worship—the cult of Cybele and Rhea, who in the Greek Islands became the great mother Aphrodite born of the sea, and in Syria Ashtaroth (Astarte). She appears also as Atargatis, the Syrian fish-goddess born from a fish's egg, and worshipped at Hierapolis; her worshippers must not eat fish. Dagon, the fish-god of the Philistines, belongs to the same group of mythologic inventions. He was half-fish and half-human, like a merman, and is, in spite of this strange personality identified with the Greek Adonis! The cult of the fish-god was widely spread in ancient Greece, even in Byzantine times, and many Christian converts were devotees of the fish worship. I have on my table a photograph of a life-sized fish modelled in gold which was dug up in 1883 from the shores of a lake near the coasts of the Black Sea. It was at one time supposed to be of mediaeval workmanship, but is now shown to be of ancient Greek workmanship (450 B.C.), and was probably a votive offering connected with the worship of the fish-god.

Then, again, in the ancient Indian story of the Deluge we read of Manu (who is the Noah of that variety of the ancient legend) finding a remarkable young fish in a stream where he is bathing. The young fish (which is really the god Vishnu in disguise) can talk, and requests Manu to take care of it, and promises him if he does so to reveal to him when the deluge is coming on. Manu takes the fish home and rears it. He then is told by the fish to prepare an ark, and place on board useful anim ls and seeds and then to embark on it with his family. The ark floats away in the flood, guided by the sagacious fish, which seizes a rope and, swimming in front of the ark, tows it to a mountain in Armenia (Ararat!), where the vessel rests whilst the flood goes down.

There was evidently a special cult of the fish in Syria and the East, which spread to Greece and Rome in very early pre-Christian times, and survives in some of the stories in the "Arabian Nights" about human beings being turned into fish. It is not surprising that this cult should have lodged itself by obscure means in the practices of the early Church.

The most remarkable outcome of this is the recognition of the fish as the symbol of Christ. The letters of the Greek name for fish ΙΧΘΥΣ (ichthus) can be interpreted as an acrostic, the component letters of the word taken in order being the first letters of the words *'Ιησοῦς Χριστὸς Θεοῦ Υἱός, Σώτηρ* (Jesous Christos Theou Uios Soter), which are in English "Jesus Christ Son of God, Saviour." This coincidence enabled the pagan worshippers of the fish-god to make their symbol or "totem" (using that word in a broad sense) the symbol of the Christian religion. Whether the use of the fish and of the letters of the Greek name for it was or was not independently started by the early Christians, its employment must have conciliated the fish-worshipping pagans, and rendered it easy to bring them into the fellowship of the Christian Church. Hence we see that a fish has more to do with Christianity than appears at first sight. It is quite possible that whilst the cult of the fish-god or fish-goddess may have involved at one period of its growth an abstention from the eating of fish or of particular species of fish as being sacred, yet the very ancient belief in "contagious magic" and the acquirement of the qualities of a man or an animal by eating his flesh, may have in the end prevailed and led to the eating of fish, the sacred symbol, on the fast days prescribed by the Church, when a special significance would be attached to such food as was sanctioned.

The evidence of the connexion of the early Christian Church with fish worship becomes convincing when once the importance of the great secret cult of the "Orpheists" and its connexion both with early Christianity and with fish worship is recognized.

It has long been known that there is a special association of the very ancient and primitive Greek cult of Orpheus, with the much later cult of Christianity. Many of the most important doctrines and practices of the widely spread secret society of the Orpheists closely resemble those of Christianity. Carvings and medals of Orpheus bringing all animals to his feet by his music were, by the earliest Christians, adopted as equally well representing Christ the Good Shepherd. But recent discoveries carry the matter much further. Orpheus is one of the names of a mythical hunter and fisherman of prehistoric times, who taught his people music, and by his magic helped them to successful catches of fish, and to the "netting" of beasts, as well as of fish. His followers adopted the fish as their "totem," or sacred animal, and they represented Orpheus (whether known by that or other names) as the warden of the fishes, a fish-god, and himself a fish—"the great fish"—and a "fisher of men.' Fishes were kept in his temples and eaten solemnly (at first in the raw condition), in order to transmit to his worshippers his powers.

In Greece, where the cult of Orpheus was introduced by way of Thrace, he became mixed with, or made a substitute for, Dionysus (the wine-god), and the same legends were told about the one as the other. He and his followers are pictured as wearing a fox's skin (supposed by some to have been originally the skin of a sea-fox or shark), and the fable of the fox and the grapes, and the very ancient story of the fox fishing with his tail, belong to the Orpheus legends.

Very ancient peoples, earlier than the Greeks of classical times, habitually adopted some animal as their totem and name-god—as do many savage races to-day.

Thus, the Myrmidones of Thessaly had the ant (myrmes) as their totem, the Arcadians the bear (arctos), the Pelasgi, who preceded the other tribes in Greece—the stork (pelargos). It is now suggested that the Hellenes, who succeeded the Pelasgi, and gave their name to Greece (Hellas) and to all its people, were so called from their having the fish (ellos, the mute or silent one, a common term applied to fish) as their "totem," and that they were, in fact, from the first worshippers of the fish-god Orpheus, Di-orphos, Dagon or Adonis! Other "cults" grew up among them. The whole Olympian company of gods and goddesses were fitted out by poets and priests with man-like forms, and with the speech, habits, and passions of humanity. But the old deep-rooted worship of the primeval fisherman who was typified by and identified with "the great fish"—much elaborated by its hymns and mystic ritual, its lore, and its legend—flourished and developed wonderfully in secret, wherever Greeks were found. Its priests were missionaries like the mendicant friars of later days, and it was — in pre-Christian times — the most popular cult not only in Greece and Asia Minor, but also in Southern Italy. Hence it is easy to understand that Christianity, by adopting the fish—the ΙΧΘΥΣ—as its emblem, readily received sympathy and converts from the Orpheists, and that the solemn rite of eating the fish on appointed days was established. Hence it seems to have come about that the early Christian Church permitted the eating of fish on most (but not on all) fast days.

Some of my readers have seen the Greek word for "a fish" stamped upon Prayer Books, or possibly a fish embroidered on the hangings of the church where they go to celebrate the birth and the passion of Christ, as their ancestors have done for a thousand years. And now they will understand the origin of the association of the sacred fish with Christian ornament, derived from a lingering pagan reverence for the mysterious silvery inhabitants of deep pools, great rivers, and the sea. It is to such survivals of the now dim rituals and celebrations of ancient days that we owe the joyful holly and the mystic mistletoe, still happily preserved in our festivities at Christmas and New Year.

The use of fish as a regular article of diet is very widely spread. Fresh fish is considered by medical men to be more easily digested than the flesh of beasts or birds, and a healthy substitute for the latter. Almost everywhere where fish are eaten, the practice of drying, and often of salting, fish, so as to store them for consumption after an abundant "catch," has grown up, and with it a great liking for the flavours produced by the special chemical changes in the fish arising from salting and drying. Ordinary putrefaction produces very powerful poisons in the flesh of fish. They are known as "ptomaines," and are produced in the flesh of fish more readily that in that of other animals. But the process of drying in the sun or of salting and smoking the fish averts the formation of these poisons. It seems, however, that a diet of dried fish is responsible for a certain kind of poisoning in man, which renders him liable to the attack of the terrible bacillus of leprosy. The leprosy bacillus must get into the body by an abrasion or crack in the skin, through contact with a person already infected. It is known that the lack of fresh vegetable and animal food produces the ulcerated unhealthy condition called "scurvy," and a "scorbutic" state of the body seems to be favourable to the establishment in it of the leprosy bacillus. The substitution of fresh meat and vegetables as a diet in place of dried fish and salted meat has apparently been one of the chief causes of the disappearance not only of "scurvy" but of leprosy from Europe. Leprosy is rapidly becoming extinct in Norway. It still survives in a few localities, and is common in several uncivilized communities in remote regions, such as parts of Africa, India, China, and the Pacific Islands. In an earlier chapter, p. 292, I have referred to the disease known as "scurvy," which has become so uncommon now as to have escaped thorough investigation by modern pathologists.

A few marine fish are known which are highly poisonous to any and every man, even when cooked and eaten in a perfectly fresh condition, and there are many individuals who suffer from the "idiosyncrasy," as it is called, of liability to be dangerously poisoned not only by the peculiar and rare fish which are poisonous to every one, but by any and every fish they may eat, or by two or three common kinds only. Thus, some persons are poisoned if they eat lobster or crab, or oysters or mussels, but can tolerate ordinary fish. Others are poisoned, without fail, by mackerel and by grey mullet, but not by sole or salmon. The symptoms resemble those produced in ordinary persons by the "ptomaines" of putrid fish, and seem to be due to the presence even in fresh fish of a kind of ptomaine which some persons cannot destroy by digestion, whilst most persons can do so. It is literally true that "What is one man's meat is another man's poison."

The use as a "relish" of the little fish, the anchovy —allied to the sprat and the herring—preserved in salt liquor in a partially decomposed state, but not undergoing the ordinary chemical change excited by the bacteria of putrescence, is remarkable and very widely spread. Anchovy sauce is made by mashing up such chemically decomposed anchovies, and is one of the very greatest and most approved of all sauces. The anchovy is a Mediterranean fish; it is taken in small numbers in sprat-nets in the English Channel and in the Dutch Zuyder Zee. So-called "Norwegian anchovies" are not anchovies, but are small sprats. When taken fresh and cooked and eaten, the anchovy has a very bitter, unpleasant flavour, which can be washed out of it by splitting the fresh fish and letting it lie in salt and water. It was this practice of washing out the bitterness which led the Mediterranean fisher-folk to discover that if left for some time in moderately strong brine the anchovy develops a wonderfully appetizing flavour, and becomes dark red in colour, whilst the liquid also becomes red. I believe that, although it would be easy to do so, it has not been ascertained whether the red colour is due to a direct action of the salt upon the blood-pigment of the fish—as is the red colour of salt beef—or whether it is due to a special red-colour-making bacterium, as is the case with salted dried cod, which is sometimes rendered unsaleable by this red growth. However that may be, the red colour of the preserved anchovy is well known, and is produced by dealers by means of artificial pigments, if not already naturally present in the salted fish as they come to market. No one would guess on tasting a really fresh bitter anchovy that it could develop the fine flavour which it does when soaked in brine to get rid of its bitterness.

Another little fish, the Bummaloh, or "Bombay duck" (Harpodon), is taken in large quantities off the West Coast of India, and is dried and used for the peculiar flavour thus developed, which is quite different from that of the anchovy. It is a deep-water fish, and is phosphorescent. The liking for the flavours developed in these fishes by various bacteria when specially treated, is similar to that which necessity and custom has developed in our attitude to cheese. Fresh cheese is difficult to obtain. Habit has ended in our preferring stale, decomposed cheese, which has developed a whole series of flavours by the action on it of special bacteria and moulds. The Roman soldiers of the first century used a small salted fish (probably enough the anchovy) to eat with their rations of bread, and such fish were usually sold with bread. Probably the small "fishes" which, together with a dozen loaves of bread, are stated to have been used in the miraculous feeding of the multitude by Christ, were salted anchovies.

Dealers in Norwegian preserved fish not only falsely call small sprats by the name "Anchovy" in order to sell them, but they have recently prepared sprats in the manner invented by French fish-curers for the preparation of the young Pilchard. The French name for young Pilchard is "Sardines," and their Italian name even in Sir Thomas Browne's time (1646) was "Sardinos." The natural fine quality of the sardine and the skilful "tinning" and flavouring of it by the French "curers" of Concarneau in Brittany, have made it celebrated throughout the world as a delicacy. The dealers in Norway sprats—for the purpose of passing off on the public a cheap, inferior kind of fish as something much better—have recently stolen the French curers' name of "Sardine," and coolly call their sprats "Sardines." The sprats thus cured are soft and inferior in quality to the true sardines, which are a less abundant and therefore more costly species of fish. The fraudulent use in this way of the name "Sardine" has been condemned by the law courts in London, but the punishment for such fraud is so small and the profit to the fraudulent dealers is so great that our French friends have to submit to the iniquity.

CHAPTER XXXVI

SCIENCE AND THE UNKNOWN

IT is a remarkable fact that although the first efforts of the founders of the Royal Society for the Promotion of Natural Knowledge, two hundred and fifty years ago, in this country, and of other such associations on the Continent, had the immediate effect of destroying a large amount of that fantastic superstition and credulity which had until then prevailed in all classes of society, and although that period marks the transition from the astounding and terrible nightmares of the Middle Ages to a happier condition when witchcraft, sorcery, and baseless imaginings concerning natural things gave place to knowledge founded on careful observation and experiment—yet the ugly baleful relic of savagery died hard, even in the most civilized communities.

In spite of all the light that has been shed upon obscure processes, and all the triumphs of the knowledge of "the order of Nature," there remains to this day in this country a surprising amount of ignorance, accompanied by blind unreasoning devotion to traditional beliefs in magic, and a love of the preposterous fancies of a barbarous past, simply because they are preposterous! "There is something in it," is a favourite phrase, and the words put by Shakespear into the mouth of the demented Hamlet, who thinks he has seen and conversed with a ghost, "There are more things in heaven and earth, Horatio, than are dreamed of in your philosophy," are gravely quoted as though they were applicable to the Horatios of to-day. We have no reason to suppose that there are more things in heaven and earth than are dreamed of in our philosophy. Those who inappropriately quote this saying as though it were proverbial wisdom are usually persons of very small knowledge, and mistake their own limitations for those of mankind in general.

The real and effective answer to all such head-shakings and airs of mystery is to demand that the reputed marvel shall be brought before us for examination. The method of the disciples of the founders of the Royal Society is not to deny or to assert possibilities. They hold it to be futile to discuss why such and such a thing should *not* exist, and still worse to conclude that it does exist, or to hold its existence to be probable, because you cannot say why it should not exist. The real question is, "Does it exist? Is it so?" And the only way of dealing with that question is to have the marvel brought before you and subjected to examination and test. "Nullius in verba!" The mere statement of dozens of witnesses merely gives you as a thing to explain or account for, not the marvel reported, but the fact that certain persons say or are reported to say that it does. What you have to examine, in the absence of the marvel itself, is, "How is it that these people make this statement?" You must inquire into the capacities and opportunities of the witnesses. There are several possible and probable answers to that inquiry. For instance, it may be that the witnesses are merely inaccurate, or are self-deceived, or deceived by the trickery or credulity of others, or are insane, or are deliberately stating what is false. Another and often the least probable answer is that the witnesses or reporters state what they do because it is the simple truth. The statements made have to be accounted for by one or other of these hypotheses or suggestions, and each suggestion as to the origin of the statements must be tested by reference to independent facts in order to dismiss or to confirm it.

The whole of what is called "modern occultism," including spiritualism, second-sight, thought transference (so-called telepathy), crystal-gazing, astrology, and such mysteries, can only be treated reasonably in the way I have mentioned. We ask for a demonstration of the occurrence of the mysterious communications or prophecies, or "raps" or "levitations," or whatever it may be. Lovers of science have never been unwilling to investigate such marvels if fairly and squarely brought before them. In the very few cases which have been submitted in this way to scientific examination, the marvel has been shown to be either childish fraud or a mere conjurer's trick, or else the facts adduced in evidence have proved to be entirely insufficient to support the conclusion that there is anything unusual at work, or beyond the experience of scientific investigators.

It is unfortunately true that most persons are quite unprepared to admit the deficiencies of their own powers of observation and of memory, and are also unaware of their own ignorance of perfectly natural occurrences which continually lead to self-deception and illusion. Moreover, the capacity for logical inference and argument is not common. The whole past and present history of what is called "the occult" is enveloped in an atmosphere of self-deception and of readiness to be deceived by others to which misplaced confidence in their own cleverness and power of detecting trickery renders many — one may almost say most — people victims. The physician who has given his life to the study of mental aberration and diseases of the mind is the only really qualified investigator of these "marvels," and no one who has closely studied what is known in the domain of mental physiology and pathology has any difficulty in understanding, and bringing into relation with large classes of established facts as to illusions and mental aberration, the "beliefs" in magic and second-sight which are here and there found flourishing at the present day, as well as the, at first sight startling, evidence of highly accomplished men who have suffered from such delusions.

Leaving aside all these more extreme cases of what we may call "challenges" to science, let me cite one or two of the more ordinary classes of cases in which science is either attacked or treated with disdain by modern wonder-mongers. It was declared by a writer in the eighteenth century that, after all, human knowledge is a very small thing, since we cannot even tell on one day what the weather is going to be on the next; still less can we control it. That remains perfectly true to-day, although by the hourly observation and record of the movements of "areas of depression" in the atmosphere and the telegraphic communication of these records from all parts of the Atlantic region of the northern hemisphere to central stations, a very important degree of accuracy in foretelling gales, and even minor changes of weather, has been reached. Side by side with this organized study of the movements of "weather" we still have the so-called "almanacs," in which, as in the days of old, certain wizards claim to foretell the weather of a year, as well as other events. It is less surprising that these wizards should find believers when one discovers that there are actually well-to-do, "half-educated" people in England who believe at this day that the delightful clever exhibitors of mechanical tricks and sleight-of-hand are really (as they usually are called) "conjurers"—that is to say, that they conjure spirits and use the "black art." Not long ago, having published my experience of the trickery of "dowsers," and the illusion known as the "divining-rod," I received a letter in which my correspondent related that, being in the coffee-room of an hotel in a country town, he was asked by a man who was there to stretch out his hand. He did so, and the man placed four coppers in a pile upon it. The man then took up an empty matchbox which happened to be on the table, and placed it over the coppers as they lay on my correspondent's hand. After an interval of three or four seconds the man lifted the matchbox, and the coppers were gone! This, which I need hardly say is one of the most common "conjuring tricks" familiar to every schoolboy, was, according to my correspondent, proof to him that the man possessed powers "not dreamed of in your philosophy," and that such powers and those of discovery by use of the divining-rod and similar occult arts are possessed by many gifted beings!

It is to be hoped that such credulity is not very common—it is difficult to form an estimate as to its prevalence, for it breaks out in different directions in different individuals. The more impudent quack remedies for various diseases have had believers amongst all classes of society—and occasionally some enthusiast bursts out with indignation in a letter to the papers, complaining that men of science or the medical profession neglect their duty to the public and refuse to examine the wonderful cure. In all these cases the cure is either a drug which is perfectly well known and practically worthless for the treatment of the disease for which it is recommended, or—as in the case of the celebrated "blue electricity" and "red electricity" (nonsensical names in themselves) sold by an Italian swindler as a cure for cancer and patronized by aristocratic ladies and the late Mr. Stead—is found to be absolutely non-existent. In this last case the liquid sold in little bottles at a high price was nothing but plain water! A more respectable case was the advocacy a few weeks ago by a correspondent in a morning paper of a common African plant (a kind of basil) as a sure destructive or warder-off of mosquitoes when grown near human habitations, and therefore a protective against malaria. Nothing could have been more emphatic than the declaration of the value of this plant by its advocate. But a few days afterwards a letter appeared from a scientific man, giving an account of careful and varied experiments, already made and published, which show that this basil, although containing in its leaves "thymol," as do some other aromatic herbs, yet neither when grown in quantity nor when crushed and spread out in a room has any effect whatever in checking the access of mosquitoes and other flies! In this case, the reputed medical marvel was to hand: it was dealt with, tested, and, as they say in the old register of the Royal Society, "was found faulty."

CHAPTER XXXVII

DIVINATION AND PALMISTRY

THE gradual passage of the race of man from the condition of "beasts that reason not" to that of "persons of understanding and reason" has been an immensely long and a very painful one. It is not yet complete—is far, indeed, from being so—even amongst the most favoured classes of the most highly civilized peoples of to-day. Just as our bodily evolution and adaptation to present conditions is incomplete and exhibits what Metchnikoff has called "disharmonies"—that is, retentions of ancestral structures now not only useless, but even positively injurious—so does the mental condition attained by civilized man (if we do not limit our observation to exceptional instances) exhibit a retention—by means of records and accepted teaching—of beliefs and tendencies which were among the first products of the blundering efforts of human reason, and have caused atrocious suffering to millions of human beings in the long process of mental development. At one time the whole race lived in a world of delusions and fantastic beliefs—the outcome of false or defective observation rather than of false logic. These false conclusions as to many subjects were inevitable as soon as man began to reason at all. It was the necessary and injurious accompaniment of the growing habit of "reasoning" by which the more fortunate races have eventually been brought, step by step, to correct conclusions and a dominant position at the present day. The progress from the almost universal prevalence of an enormous system of preposterous false beliefs or conclusions onward to the triumph of sound knowledge has not only taken an immense period of time, but left whole races of men and large sections of the population—even in those races which have produced individuals remarkable for their power of discovering the truth—still subject to the early erroneous conceptions of natural processes and of man's relation to them.

The conclusion certainly seems to be justified that the most advanced animal progenitors of mankind, who lived and died unreasoning, the mere puppets of natural forces which they neither could, nor tried to, understand and control, were "happier" than the "rebel" man when he first conceived the notion that he could detect cause and effect, not only as between a blow and the production of a serviceable flint implement, but in the beneficent or injurious relations of the things around him to one another and to himself. Primitive men seem at a very remote period to have elaborated in regard to such vital matters a series of conclusions—differing in various races according to place and circumstance—to which they were led by erroneous observation and imperfect reasoning—reasoning which was arrested and distorted by fear, desire, haste, and imagination. The word "magic" is now used to indicate those beliefs and conclusions in all their variety, because the "magi" or priests of Zoroaster (Zarathustra), the founder of the religion of the ancient Persians, taught them in an elaborated form, and practised a system of supposed control of natural forces and of spirits, good and evil, in connexion with such beliefs. Magic is, therefore, defined as the general term for the practice and power of wonder-working as dependent on the employment of supposed supernatural or "occult" agencies. It forms a vast field of study and one of the greatest interest in the attempt to follow out the history of the workings of the human mind, its extraordinary envelopment in error and delusion, and its gradual emancipation therefrom.

In origin "magic" and "religion" are one. The priest and the magician were originally one. Man tried to control Nature by the use of spells and fantastic procedures, based on imagined powers and correspondences in natural objects. He excogitated (as a modern child sometimes does) a sort of fancifully assumed system of fixed laws of natural relations and interactions, of causes and effects which were suggested by superficial likenesses and wild guesses at connexion and sequence, accepted without criticism. Thus, we have the widespread doctrines of "sympathetic magic" and of "contagious magic." An example of the first is the belief that a certain tree or animal is the sympathetic representative of a certain man, and that as the one flourishes or suffers and dies so will the other. This is extended into a belief that a drawing or image, or even an unshaped stone, may sympathetically represent a man or an animal. The American medicine-man draws the picture of a deer on a piece of bark, and expects that shooting at it will cause him to kill a real deer the next day. He mistakes a connexion which exists only in the mind of the sorcerer for a real bond independent of the human mind. Thus, too, waxen or clay images of an enemy are made and melted before fire or wasted in water, or pierced with pins (even at this day in Scotland, as witness a clay figure in the museum at Oxford), in the belief that the enemy himself will be similarly injured.

The belief in "contagious magic" leads to the procuring of a drop of the blood, or of a piece of the hair, the toe-nails, the clothing, or even a part of the unconsumed food of another individual, in order that a sorcerer may, by acting upon it or repeating "incantations" over it, influence the actions and life of that individual for good or for ill.

But besides the many forms of these two kinds of magic, there is a later variety of magic which grew up with what is not a primitive belief, namely, the belief in the existence of spiritual beings inhabiting trees, rocks, waters, and animals. It developed further with the later belief in the existence of ghosts or spirits of the dead. Fear and the desire to control hostile unseen forces was the motive of all magic. The magician invented "spells," "rites," and "ceremonies" for controlling and bending these spirits to his will. But as a still later development, we find more and more definitely separated from the magician and his spells—the priest, who learnt humility in the face of might greater than his own, and, abandoning the attempt to coerce, adopted the attitude of propitiation and prayer, and prostrated himself before a higher power. Thus (as Dr. Marret writes) religion gradually became separated from magic, though often mixed with it, and often retaining magical elements. Religious cults became publicly recognized, established, and respectable, whilst "magic" became private, secret, disreputable, and at last openly condemned and suppressed by the priests of religion. The history of magic in Europe, Asia, Africa, and America presents an almost unlimited field of study. We find remarkable agreements in the fundamental notions on which magic is based in all parts of the world and also important differences in details and special developments.

Divination is that branch of magic which attempts

to *discover* secrets or to foresee events, whilst magic in general is an attempt to *influence* the course of events. Divination is the process of attempting to obtain knowledge of secret or future things by means of oracles, omens, or astrology. One of its methods is "necromancy," the supposed communication with the spirits of the dead. This word is formed from the Greek words "nekros," a corpse, and "manteia," divination; but in Latin it was erroneously written "nigromantia," and so gave rise to the application of the name "the black art" to sorcery and witchcraft in general. By the ancient Greeks and Romans all omens, as well as oracles, were regarded as sent by the gods, and in ancient Rome a large and wealthy corporation of augurs who were constantly consulted by private individuals as well as by the State existed. They received regular "fees" for their services in interpreting and seeking for omens. The orthodox belief has always been either that the soothsayer is directly controlled by a god or a spirit, or, on the other hand, that the material objects inspected and regarded as signs of the future are controlled by the gods or by spirits, so as to afford information. Divination is, and has been, practised in all grades of civilization and culture, from the Australian "black fellow" to the American medium. Amongst its many varieties are (1) crystal gazing, a method similar to that of dreams, excepting that the vision is set up voluntarily by gazing into a crystal ball or a basin of water; (2) shell-hearing; (3) the divining-rod in its various forms; (4) sieve, ring, and Bible swinging; (5) automatic writing; (6) sand divination, widely practised in Africa; (7) trance-speaking; (8) the examination of the hand, or palmistry; (9) card-laying; (10) the interpretation of dreams; (11) the casting of lots, or sortilege; (12) the drawing of texts from the Bible or from Virgil (the 'sortes Virgilianæ' of old times); (13) the inspection of the entrails of animals freshly killed (haruspication), and the study of footprints; (14) augury by omens, such as the behaviour and cry of birds, and the meeting with ominous animals; and lastly (15 and 16), the two highly elaborated and pretentious systems of astrology (divination by the stars) and geomancy (divination by the lie of hills and rivers). In the case of astrology the stars are believed not merely to prognosticate the future, but also to influence it, and the latter is the special feature of geomancy, practised in China, where no house or other building can be erected without a certificate as to its favourable position in regard to "magic" by the professional "geomancer," who has to be paid his fee, and thus takes the place of the local government surveyor and sanitary officer of Western Europe.

In the exercise of these arts of divination there is no doubt that, owing to the concentration of his attention on the thing to be inspected the operator is, in many kinds of divination, "self-hypnotized," or brought into that well-known mental condition in which the unconscious memory and other special mental processes are active, whilst an exaggerated acuteness of the senses is produced. In other cases the person who consults the "operator" may be so influenced. Hallucination of one kind and another is therefore likely to occur, and thus mystery and apparently marvellous results are not inconsistent with the good faith of the operator. But there is no reason to doubt that the modern sorcerers who make money by their pretended divinations are rogues and impostors of a particularly dangerous and injurious variety.

Palmistry or chiromancy is one of the oldest of the large family of systems for foretelling the future. It existed in China 4000 years ago, and is treated in the most ancient Greek writings as a well-known belief. The gipsies probably brought it with them from India. Those who practise palmistry pretend that by the inspection and proper interpretation of the various irregularities and flexion-folds of the skin of the hand the mental or moral dispositions and powers of an individual can be discovered, and not only that, but that the current of future events in the life of an individual are indicated by them. To this it is customary to add nowadays the pretence of a revelation by these same markings of events in the past life of their owner. It is only what we might have expected that primitive man, seeking for signs and occult mysteries, should have found in the varying folds of the hand—" the organ of organs " —something to excite his tendency to attribute magical importance to what he could not simply explain. The folds of the skin on the palmar surface of the hand are, as a matter of fact, so disposed that the thick loose skin shall be capable of bending in grasping, whilst it is held down to the skeleton of the hand by fibrous lines of attachment, so as to prevent its slipping and the consequent insecurity of grip. The swellings bounded by the lines of folding and fixture are called " monticuli " by the palmist, and are simply subcutaneous fat, which acts as a padding, or cushioning, and projects between the lines of fibrous attachment of the skin to the deeply placed bones. They differ slightly in different individuals, as do other structures.

These same lines and monticuli are present in the hands and feet of the chimpanzee and other man-like apes, and were specially exhibited under my direction in the upper gallery of the Natural History Museum. But no palmist ever read the ape's hand, although, according to the great and authoritative treatises on palmistry, it would be perfectly easy to do so, since every variation in the lines and the monticules has been mechanically dealt with, and its supposed indications precisely determined by a formal set of rules. There are similar lines on that part of the foot in human infants and in the adult apes which corresponds to the palmar surface. But no palmist has attempted to deal with them. The fact is that the attributions indicated by such names as the line of heart, the line of life, the line of the head, and the line of fortune are purely arbitrary, as are those of the monticules Venus, Jupiter, Saturn, the Sun, Mercury, Mars, and the Moon. In past times there have been great divergences in their interpretation by different schools, and the present uniformity is as devoid of any conceivable relation to fact as were the former divergences. It is impossible to discuss the asserted correlation of the lines and monticules of the hand with either character or life-history, since no facts are offered in support of the notion that there is such a correlation. We have bare assertion, and nothing more, as in most of the other doctrines of magic.

The shape of the hand and of the fingers, and the softness, hardness, dryness, and moisture of the skin are taken into account by most palmists. Few, if any, of those who pretend at the present day to "read" a hand are really acquainted with the elaborate rules laid down by the painstaking, if deluded, people who endeavoured to construct a sort of astrology of the hand by assigning the names of heavenly bodies to parts of it. The modern professional palmist forms a judgment and guess as to his or her client's character and probable past and future history by indications and information obtained from the client's face, manner, conversation, costume, and personal acquaintance. If a vague prophecy made by the "fortune-teller" should by hazard turn out to be near the truth, it is remembered and quoted by the client as a proof of the truth of palmistry; if it does not prove to be correct, it is forgotten.

The question of the possibility of judging of the character and disposition of a man or woman by the form and proportions of the hand or the foot is altogether distinct from that of the reality of "divination" of future events by applying a system of rules to the interpretation of the lines and swellings of the palmar surface. Persons of quick perception are in the habit of forming judgments as to character from a first impression of the face, expression, voice, and movements of another individual. Often such judgments are erroneous, and I do not know that they have ever been proved by a large series of experiments to be more frequently right than wrong. But it is possible that correct indications may sometimes be thus obtained. Many people think that they can form more or less correct judgments as to certain mental characteristics by observing the shape and play of the hand and fingers or of the foot. There may be such a correlation of the gesture and form of hands or feet with some mental qualities, but obviously this has nothing to do with palmistry. It has never been really proved that persons of what is called "good birth" have smaller hands and feet than persons of "low birth," although it is often assumed that they have. And it has never been shown why small hands and feet should go with "good birth," supposing that they do so, or why some people have large and some small extremities. The possible effect of certain manual occupations in enlarging the hands of an individual is, of course, excluded; the question raised is as to naturally or hereditarily small hands and feet.

CHAPTER XXXVIII

TOADS FOUND LIVING IN STONE

IT is quite true that one should not refuse to entertain the possibility of something almost incredible taking place, simply because it is highly improbable that it has taken place. Also it is important that one should not accept and believe in the reality of the marvellous occurrence, merely because a decent sort of person has asserted that he has witnessed it and is satisfied of its reality. In a previous chapter (p. 117) we have seen how the story of the Tree goose and the hatching of geese from Barnacles was supported by respectable but incompetent witnesses such as Gerard, the herbalist, and Sir Robert Moray, the first president of the Royal Society. There are many equally baseless fancies which are attested by "respectable" witnesses at the present day.

The statement that workmen splitting large blocks of stone in the quarries have seen a toad hop out of a cavity in the interior of the stone attracted a good deal of attention in the earlier half of last century. I do not know whether it can be traced to any great antiquity. I see no reason to doubt the truth of the statement in its simple form as given above. It has, I have no doubt, repeatedly happened—as letters to newspapers and in earlier days serious pamphlets record

—that on splitting a block of stone the workmen engaged in the operation have seen a toad emerge from the broken mass. The fact is that the rocks in many stone quarries are "fissured" or cracked, so that a narrow space or "crack" extends through many feet of thickness of rock to the surface, which is covered by vegetable mould. Occasionally, owing to rain and flood, the mould is washed away, and some of it carried into the cracks or fissures in the rock. Occasionally a young toad is carried from the surface into such a fissure and far down its sides, and eventually lodges 20 feet or more in the thickness of the rock. The same circumstances which have carried the toad into the fissure carry in also from time to time small worms, grubs, insects, on which the toad may feed, but in any case the far-spreading though narrow fissure will hold plenty of air and moisture, and even without food a toad can remain alive for several months provided that the temperature is about that of a cool autumn day, its surface kept moist and the air also. Hence it is in accordance with recognized conditions that occasionally quarrymen should "get out" a block of stone deep below the surface in a stone quarry which is traversed by a fissure or has a small natural cavity in it (as limestone and other rocks often have) communicating with a fissure, and that when they break the stone and accidentally open the fissure or connected cavity a healthy living toad is found ensconced in it. The recent washing of clay and powdered stone into the fissure by rain and flood sometimes may hide its existence from the casual observation of the workmen, and the soft material washed in may even be found fitting closely to the toad's body. And thus it will appear that the toad is very closely embedded in the solid stone.

Probably no one would have cared very much a couple of hundred years ago if toads were constantly present in the centre of solid stones. Toads were regarded as queer, dangerous things connected with witchcraft, and there was no accounting for their behaviour. The view taken by the well-to-do class would have been in those days (as perhaps it would be less generally to-day) similar to that of the Chicago millionaire when shown, by means of the spectroscopic examination of light, the proof of the existence of the metal sodium in the sun. The professor who took the millionaire round his laboratory wished to interest him in the discoveries of science, and hoped that he might contribute to the funds necessary to pay for the elaborate and delicate instruments by which such discoveries are made. He showed many remarkable experiments to his visitor, and wound up by showing him the two narrow lines of yellow light caused by incandescent sodium. He showed him how exactly their position in the spectrum could be fixed and measured; how they caused two black lines in the spectrum of light, which was made to traverse a flame in which incandescent sodium was present. And then he showed him that in the spectrum of the sun's light there were two black lines (besides thousands of others) which exactly coincide with the two sodium lines; whilst others of the black lines in the solar spectrum coincide with bright lines given out by incandescent hydrogen, iron, magnesium, etc. The millionaire followed it all and understood the completeness of the demonstration. The professor was delighted and hopeful. Then the millionaire said, "Who the hell cares if there is sodium in the sun?" I was not told by the disappointed professor (it was Professor Michelson, and he related this little episode at a dinner of the Royal Society) what reply he made to this inquiry or whether he was eventually successful in his attempt to secure funds from the millionaire. The attitude which the millionaire took towards scientific discovery is not a natural one, but the result of the stifling of natural interest and curiosity by long concentration on the art and practice of money-making. So, too—owing to other mental pre-occupations and concentrations—though a boy or a savage might have been puzzled and deeply interested in the occurrence of a live toad in the middle of an apparently solid piece of rock, the "country gentleman" of the eighteenth century would have said, if the matter had been pressed on his attention, "Who the hell cares if there are live toads in the rocks?" And a large but decreasing number of his representatives to-day would make the same remark.

It, however, happened that at the beginning of the nineteenth century a spirit of inquiry into the history of the crust of our earth was set going. The science of geology was eagerly pursued by many capable men, both abroad and in this country. The Geological Society of London was founded in 1809. The doctrine of the vast age of the earth and the demonstration of successive layers of deposit—forming its rocks and containing the remains of strange and of gigantic animals unlike those now existing—excited widespread interest and controversy. Buckland introduced the study of geology in Oxford. Lyell was his pupil, and became the great teacher and exponent of geological theory in a series of masterly treatises, written in such form that they appealed during half a century to educated men of all professions and occupations. The country clergy and their friends gave themselves with enthusiasm to the investigation of strata and the collection of fossils. Now came the opportunity of the toad embedded in stone!

It is not worth while inquiring who was the first to make the suggestion, but it very soon became one of the favourite assertions of the wonder-mongers who hang on to the skirts of science—not to be confused with the enthusiastic nature-lover—that the living toads found in blocks of stone, and sometimes in lumps of coal, are thousands of years old, contemporary with the geologic age of the rocks in which they are found embedded, survivors of the extinct animals whose bones and teeth the geologists had discovered and described, also embedded in such rocks! This entirely baseless fancy took root, and has flourished ever since the early Victorian period. Only a few months ago there were paragraphs in the papers on the discovery of a live toad of antediluvian age in a block of stone. Old gentlemen have repeatedly written to the newspapers, and sometimes privately to me, describing how they had, on breaking an unusually large lump of coal in the dining-room coal-scuttle, liberated from an age-long prison an antediluvian toad, which hopped out from the lump of coal in a marvellous state of health and agility. Whenever any discussion has arisen with regard to these statements, and such an explanation offered as I have given above as to the apparent enclosure of a toad in a piece of rock, or a similar explanation as to the encasement of one in the black mud adhering to lumps of coal stacked in sheds or cellars—some of the would-be believers in the immense age of the liberated toads appeal to the fact that amongst the most remarkable extinct animals whose bodies are found in ancient strata are reptiles, whilst others, more learned, insist on the well-known prevalence of the remains of animals of the class Amphibia, to which the toad belongs, in the "Coal Measures."

The answer to these rash believers in what they

call "the evidence of their own senses" and the disentombment of living specimens of the ancient world from lumps of stone or of coal—apart from that given by the fact that there is complete absence of any proof that the toad before liberation was really and truly encased in a stony chamber to which it could not, by any possibility, have recently gained access—is that the common toad, which is thus discovered and supposed to be a survivor of long past geologic ages, is a modern production of Nature's great breeding establishment. It is quite easy to distinguish it from all other living species of toads; it is spread over a limited area, existing in the north temperate region of our hemisphere in many parts of which it is replaced by other similar but distinct species. If we ask what is known of it in past ages as revealed by the Pliocene, Miocene, and Eocene strata, we find that it did not exist at all in the latest of these, but was represented by ancestors like it, yet markedly different. Remains of a kind of toad are found in the Upper Eocene "phosphorite" of the South of France, and in 1903 such remains were found in an oolitic deposit. As we descend further the series of geologic strata, the remains of toads and frogs cease to occur. In the coal measures they were represented by ancestors provided with tails like the newts and salamanders of our own day. They had not come into existence, nor, probably, had any creature closely resembling them, at that period. In the "Coal Measures" we find abundant remains of very large and also of small animals related to salamanders, newts, and less closely to toads, but they are in great and important features of structure unlike the Amphibia and Batrachia of to-day. Hence the notion which lay at the bottom of the excitement caused by the discovery of live toads in the interior of rocks or of coal—namely, that the creature was a survivor from the lost world of extinct "antediluvian" animals—falls to the ground. It has no better claim to attention than the similar but perhaps bolder statement indulged in from time to time by an inventive transatlantic Press, namely, that "some workmen on blasting a rock in the quarries at Barnumsville were astonished by the escape from a cavity within the solid rock of a large flying lizard or pterodactyle which immediately spread its wings and flew out of sight."

Connected with these fancies is the theory that the traditional dragon of heraldry and of the Chinese is a memory handed down to the present day from immensely remote times, when—so we are asked to believe—man co-existed with the great extinct dragon-like creatures known as pterodactyles (see "Science from an Easy Chair," First Series; Methuen, 1910). As a matter of fact the heraldic dragon does not closely resemble the pterodactyle or other extinct reptiles, and is an imaginative creation of human artists based upon the realities of the great pythons of India and the little parachute lizard (8 inches long) of the same region, known to zoologists as Draco volans. The close agreement of this little lizard with European heraldic representations of the dragon is conclusive as to the origin of the details of form and appearance assigned to that legendary beast, though the great size ascribed to it and the terror associated with it is traceable to the great snakes of the Far East—"drako" being the Greek word for a serpent. And further, there is very good ground for concluding that a long interval of geologic ages separates the disappearance of the great extinct reptiles and the pterodactyles from the appearance, on this globe, of the earliest man-like apes, and no reason to suppose that the latter could have handed on any knowledge of such extinct reptiles to their descendants, even had they seen such creatures.

CHAPTER XXXIX

THE DIVINING-ROD

THE divining-rod, spoken of by the Romans as "virgula divina," and mentioned by Cicero and by Tacitus, was a different thing altogether from the modern forked twig of the water-finder, and seems to be of immemorial antiquity. Its use in "divination" was similar to that practised with a ring or a sieve suspended by a string. When the rod is thrown into the air and falls to the ground, or when the suspended object is set moving, it eventually comes to rest, and when thus at rest must point in one particular direction. It was supposed that gods or spirits invoked at the moment guided the movement and final position of rest, so as to make the divining-rod or ring or sieve point to buried treasure, to an undetected murderer, or to a witch or wizard who had used magic arts to injure the person seeking its aid. Bits of stick are so used at the present day by some savage races. The notion leading to its use is the same as that which has led to augury by inspection of an animal's entrails, by the flight of birds, and other such varying appearances. The notion is that an unseen protective power will, when properly invoked, interfere with the blindly varying thing and make it vary so as to give indications either of hidden objects or of future events. The unseen power which thus revealed itself was primitively supposed to be that of a god or a spirit, but later the augur or intermediary who worked the "show" acquired exclusive importance and arrogated to himself mysterious powers. The same transference of importance has come about in the case of the modern hazel-twig and the "douser," who now claims to "divine" without its aid.

The tossing of a halfpenny to decide as to alternative courses of action, still almost universally prevalent in this country, is in origin (and largely in actual practice) an appeal to supernatural powers to give an indication by interference with the natural fall of the coin, as to which of the alternative courses is the more favourable to the interests of the individual who tosses the coin or agrees to follow its decision if tossed by someone else. "Heads I go; tails I stay where I am." Of a like nature is the drawing of lots, and so are a number of similar practices originally devised for the purpose of obtaining guidance from supernatural sources. Some of them have survived without any associated superstition, and are commonly used at the present day merely in order to obtain an impersonal decision as to which of two or more claimants is to enjoy a certain privilege or exemption, as, for instance, when a coin is tossed to decide as to which side of the river at the start shall be occupied by competitors in a boat race, or which shall have choice of innings in a cricket match, or as when lots are drawn to determine who shall enjoy exemption from military service. But even in these cases there are large numbers of men and women who believe that some mysterious power which could possibly be won over to their side, or else what they call "a special providence," determines the issue. There are, I need hardly say, no facts which justify the belief in any such interruption of the orderly course of nature.

The forked twig (virgula furcata of the alchemists) used by water-finders has another significance and history. The forked twig is held, one branch in one hand and the other branch in the other hand, by the explorer. After a time, as the explorer walks along, the twig suddenly, and even vigorously, " plunges " or " ducks " as he holds it. It seems to do so " of its own accord." The old English word " douse " signifies ducking, dipping, or plunging. The forked twig " douses." Hence the persons who use it are called " dousers." The belief is widespread that this dousing or plunging of the forked twig is caused by the presence of a vein of metallic ore in the ground, or in other cases by the presence of subterranean water. It is interesting to ascertain what grounds there are for this belief.

The dousing-rod or twig is first mentioned in the fifteenth century by a writer on alchemy (Basil Valentine), and in 1546 by Agricola (De re metallica), who says it must be either of willow or hazel, and describes its use in the discovery of metalliferous veins and subterranean water. The purely fantastic belief on which its use was based was part of the doctrine of " sympathies." It was supposed that the branches of certain plants were drawn to certain " sympathetic " metals in the earth beneath them—a supposition suggested by the downward growth or " weeping " of the branches of trees and bushes in some cases. By the Germans the forked twig used in searching for metals or water was called " Schlagruthe," which has the same meaning as " dousing " or " plunging " or " striking rod." It was introduced into England by German miners who were employed in the time of Queen Elizabeth by merchant venturers in working the Cornish mines—and it has remained with us ever since—though one hears little at the present day of its use in searching for metalliferous deposits, and more about the supposed wonderful results obtained with its aid by professional water-finders.

We have to distinguish the facts established in regard to "the dousing-twig" from the inferences and suppositions based upon those facts by credulous people. There is no room for doubt that when the forked twig, in shape like a letter Y upside down, is held by a more or less nervous but perfectly honest person who takes the matter very seriously, and holds firmly one branch of the fork in one hand and the other in the other hand, the fingers well round it so as to bring it against the palm of the hand, a strange thing happens after some minutes. The twig seems to the person holding it to give a sudden movement as though drawn downwards. If he or she is walking along, intently awaiting this movement, and believing that it will be caused by some subterranean attraction, the effect is, naturally enough, startling. It occurs more readily with some persons than with others. What is the explanation of it? There is no necessity for supposing that it is due to any mysterious attraction by hidden water or metal. It has been clearly shown that it is due to fatigue of the muscles which are employed in keeping the hands and fingers in position. The muscles in use suddenly relax, and the hands turn to a new pose—one of rest—and with them the forked twig. In most persons attention and control are sufficiently active to prevent this sudden relaxation of the muscles. But those who are liable to mental absorption in the strange procedure, and are apt to become half-dazed by the solemn sort of "rite" in which they are engaged, find their tired hands (tired, though they are unconscious of it) suddenly turning, and the twig "ducking" downwards in a way which they can neither explain nor control. Such persons are the honest, self-deceived "dousers," who are, and have been, sufficiently numerous to establish a belief in the existence of a mysterious agency causing the twig to "duck." No doubt originally, with complete innocence and honesty, this mysterious agency was believed to be a sort of magnetic attraction due to a sympathy between the twig and subterranean metal. In later days, without any attempt to give a reason for the change, the same class of people have believed that it was water far below the surface of the earth which was the cause of the attraction, and consequent ducking or dousing of the twig.

Let us assume for a moment that the facts are as I have stated, and that the honest "douser" merely finds his forked twig dousing or ducking because his hands are tired by keeping in one position. Then it is evident that no harm would be done, but rather a useful decision leading to action would be determined, by the belief that concealed metal was the cause of the "ducking." Digging must be commenced somewhere, and the dousing-rod would only be tried on likely ground, so that, often, the thing sought for (whether metal or water) would be found after prolonged excavation at the spot indicated by the douser, or near it. If the digging were a failure, the believers in the dousing-rod would say that they had not been able to dig deep enough, or that some hostile agency had intervened and misled the "douser," or that he was in poor health, and so "worked" badly. The successes are remembered and the failures forgotten. So the belief in the dousing-twig as a real guide to subterranean metal and water has been maintained, and all the more securely because there have been, and doubtless are still, many honest, innocent country people who truly believe that they possess an exceptional and mysterious gift in being able to experience the curious ducking action of the twig when they walk with it in their hands in quest of this or that.

In the seventeenth century the dousing-twig was used as a guide in all sorts of quests, for instance, in searching for hidden treasure and in tracking criminals! In our own times it is chiefly known through its use by professional water-finders. There is no doubt that some of these gentry are dishonest. They are not the credulous rustics to whom the dousing-twig owes its long popularity. They are often clever and expert judges of the indications by form of the land, lie of geological strata, and distribution of vegetation, as to the subterranean water which is so abundant in this country. They make a pretence of using the douser's twig, in order to obtain employment from landowners in search of a likely spot for sinking a well, since it is the fact that many people prefer to be guided by a sort of magician who uses a supposed mysterious occult agency rather than to employ the honest and perhaps less acute geologist who avowedly proceeds in his search for water by making use of ascertained facts as to the structure and character of the subsoil and deeper strata of the district in which his services are called for.

The believers in the connexion of the movement of the douser's rod and the existence of concealed metal or water, have of late years started the theory that the twig itself is of no value in the "experiment." Certain dousers have declared that they can work just as well without it, and that it is not the rod or twig but they themselves who are sensitive to concealed water or metal. They state that they feel a peculiar "sinking" in the pit of the stomach, also a nervous tremor, and that their hands move spasmodically, causing the rod to move, and they attribute this to an influence on the human body of "vibrations" or possibly "electricity" from the concealed metal or water. This is ingenious enough; it shifts the seat of mysterious action from the simple twig to the much more complex human body, and accepts to a certain extent what I have above stated as to the nervous condition of the douser and the fatigue of the hands.

Others, who have lately discussed the subject, suggest that the douser is affected not by any known kind of physical vibrations, but by some mysterious emanation from the concealed metals or water similar to that which they (without any sufficient evidence) assume to pass from one human being to another over long distances, causing what has been called "second-sight," "thought-reading," and (in order to give an air of scientific importance to it) "telepathy." This may seem satisfactory to some people, but it is plainly a case of attempting to explain a little-known thing by reference to a still less known thing—what is called "ignotum per ignotius." Sir W. F. Barrett, of Dublin, has lately written on this subject, and very rightly says that the real question to be decided in the first instance is whether the modern "water-finders," who profess to be guided by occult influences, whatever the nature of those influences may be, are more successful in discovering water than those who seek for it by the use of the known natural indications of its presence; and, further,—and this seems to me to be the most important consideration,—whether, taking into account all the "experiments" made by the occultist water-finders, both the successful and the unsuccessful, the proportion of successes is greater than might be expected as a matter of chance and the use of common intelligence.

That is, in fact, the interesting point about the persistent belief in the "magical" powers of water-finders. It is one of several more or less traditional beliefs which depend on coincidence. The belief in birth-marks is of this nature. A lizard drops from the ceiling of her room on to a woman. A few weeks afterwards she bears a child which has a mark upon its breast more or less "resembling" a lizard. Some people believe that the mark on the child is caused by what is called "a maternal impression," the influence on the mother's mind of the scare caused by the lizard being expressed in the mark on the child's body. To form a conclusion as to the truth of this explanation we require to know what proportion of mothers in a given population have been startled by lizards, what proportion of children are born with marks on them more or less "resembling" a lizard (there is much significance in the "more or less"), and whether there are more children born with a lizard-like mark on the body from mothers who have been frightened shortly before the child's birth by a lizard, than from mothers who have not been thus frightened. The inquiry is not an easy one. The same question of coincidence applies to water-finding. Taking several thousand attempts to find water we must ask, "Is the attempt unsuccessful in a larger percentage of trials in the case of those who do not follow the indications of a dousing-rod than in the case of those who make use of it?" Sir W. F. Barrett admits the difficulty of getting at satisfactory statistics in the matter; but is inclined to think the dousers are the more successful, and so entertains a theory of mysterious agency to account for their success. My own impression is that in difficult cases of search for water dousers are as frequently unsuccessful as non-dousers.

It is true we cannot get proper returns of all cases of success and failure. But in this matter of "water-finding" we can make use of "experiment," a thing which is not so easy in regard to birth-marks—though it is related that the patriarch Jacob made an experiment of this character with his pealed stakes. Experiments have lately been made with dousers or water-diviners to test their powers. These experiments have been carried out both in Paris and in the South of England. They are unfavourable to the pretensions of the diviners.

It is very difficult to perform under perfectly fair conditions a number of experiments sufficiently large to enable us to arrive at a demonstration of the truth in this matter. Some thousand "dousers" should be put to the test under proper conditions and guarantees, and the percentage of failures and successes carefully recorded. This has not been done, although "dousers" have often been tested and found to be unable to discover subterranean water known to be present, or else have given erroneous indications. If you prove some one individual "douser" to be an impostor, or else self-deluded—the reply by those who believe in the existence of the occult power attributed to dousers is, naturally enough, that though this individual was an impostor, or incapable, yet that does not prove that all other individuals who claim to possess certain peculiar powers in the discovery of water are so. All that can be done is to challenge any douser to come forward and establish, in the presence of a competent tribunal of experts, that he can indicate in a given area the whereabouts of subterranean water already known to the committee but not possibly known beforehand to the douser.

This experiment was made a year or two ago near

Guildford by a committee of water engineers and geologists, and also by a similar committee in Paris. Only a dozen or two of the water-finding dousers came forward and submitted to be thus tested, and they entirely failed to show any special capacity for discovering water. They failed signally. But then the believers may, of course, retort that the really gifted superior dousers had refused to have anything to do with the inquiry, and that "their withers are unwrung." The same kind of test was some years ago made with the so-called "spiritualist mediums." A banknote for £1000 was placed in a very carefully sealed envelope, and deposited in a safe in a bank. Its owner advertised his offer to present the note to any spiritualists who would correctly state the number of the note. The offer remained open for some years, but the spiritualists were unable to gain information about this very simple matter by their methods of consulting supposed "spirits," and the note was never claimed. Of course, some of those who believe in spiritualism, maintain that the genuine "mediums," for some reason not altogether clear, refused to make the attempt to discover the number. Others put forward the view that the "spirits" took offence at the proposed test, and refused to reveal the number. Others, again, took the line that this was just one of the few things about which "spirits" are unable to communicate with mortals, or are forbidden by superior order to reveal.

It is accordingly fairly obvious that it is not of much use to take the trouble to expose the falsity of the pretensions of any isolated specimen of a douser or of a spirit medium. However that may be, some years ago, when I was staying in an ancient castle in the North of England, my hostess procured the attendance of a youth who had a great reputation as a douser, in order that I might test his pretensions. The youth arrived with his father, and had half a dozen Y-shaped hazel twigs ready for use. The party staying in the castle met him on the terrace, a broad gravel walk which surrounded the battlements. I asked him to walk round the castle and mark in our presence the spots at which his twig indicated the presence of subterranean water. The circuit was somewhat less than a quarter of a mile, and he indicated eleven spots. We placed obvious marks at each of these spots. I then took him into the castle and, aided by a friend, carefully blindfolded him with pads of cotton-wool over each orbit and a large silk handkerchief. We then led him out by a circuitous route on to the terrace and asked him to try again to indicate the spots which he had just discovered. He walked along as before and stopped at several spots, saying that his twig indicated water where he stood. He also made futile efforts by turning and throwing back his head, to catch a glimpse of some of the marks we had placed at the spots previously indicated by him. But the pads of cotton-wool effectually prevented him from seeing anything. In no case (as a large party of onlookers testified) were the spots indicated on his second circuit identical with, or even near to, those marked in the first circuit. His father said he was "upset" by the blindfolding. We then removed the bandage, and took him into a large courtyard beneath and across which from one corner to another a large subterranean conduit ran. We had arranged that the water should be running in abundance through this conduit. We told him that such a subterranean channel existed. He was left free and undisturbed, and his eyes were not bandaged. But he failed to discover the conduit altogether, although he crossed it several times; and he ended by declaring that his twig indicated subterranean water at a spot remote from the conduit, where some large vats stood for the purpose of storing rain-water! All this, of course, tended to prove the incompetence of the youth as a douser, and to make it probable that such successes as he had obtained elsewhere (and my hostess stated that they were very numerous and remarkable, and vouched for by members of her own family) were due to imposture.

But a single case like this does not bring one very far on the way to deciding the question as to whether there are persons who are genuinely and successfully guided to the discovery of subterranean water by strange sensations and by spasmodic movements of their limbs or of hazel-twigs held in the hands, due (as they declare) to an obscure influence which emanates from subterranean water and from buried metal. The fact is that we have in the belief in the guidance of the douser by occult influences a troublesome case of the fallacy in reasoning expressed by the words, "post hoc ergo propter hoc," or, to put it in English, "after this, therefore caused by this." Primitive man found that this mode of forming a conclusion very often led to a correct discovery of the connexion between two events, and he adopted it as a ready method of guidance, although it was frequently fallacious. It has taken ages, literally ages, to make people discard this mode of arriving at a conclusion in serious matters, and it is still usual in less vital affairs. To show that B followed upon the occurrence of A, even once, is, of course, a proper and useful way of forming a guess or a suggestion as to the cause of B, but still more is your guess legitimate if the sequence has occurred several times in your experience. But it is only a guess: a conclusion must not be accepted on that basis, although lazy and hasty people do adopt such conclusions. You must find out the details of the nature of A and also of B, and if possible how the one is connected with the other. And if you cannot do that you can still establish your conclusion and confirm your guess by showing that B *invariably* follows upon A, or that (in a long experience) only when A has been present, and never when A has not been present, has B occurred. If you cannot prove the truth of your guess by this experimental demonstration of the exclusion of other causes than A or by the experimental demonstration of the invariable occurrence of B after A has occurred, then you have to seek for evidence of a real connexion between A and B, though not an invariable one, by collecting a vast number of instances of the occurrence of B and finding out whether A has preceded it in such a large proportion of cases (as compared with those in which B has occurred without the previous occurrence of A) that the cases in which B follows A cannot be considered as accidental, but indicate a real causal relation of A to B.

This is always a difficult undertaking, whether we start with the guess that B is caused by A or that it is not caused by A. In the case of water-finding, water is found at depths of 30 feet to 100 feet and more below the surface by engineers without the aid of "dousers" every day, and this is so frequent and regular a proceeding that the percentage of cases in which dousers find water, that is to say in which B—the discovery of water—follows A (A being the employment of a supposed sensitive douser with or without his twig) does not—so far as I am able to judge without strict statistical evidence—exceed the percentage of successes in searching and digging for water by ordinary intelligent men without the introduction of A.

CHAPTER XL

BIRTH-MARKS AND TELEGONY

TWO widely-spread "beliefs"—in regard to the complicated and not generally familiar subject of the reproduction of animals—are, in addition to that dealt with in the last chapter, examples of the unjustified and primitive mode of forming a conclusion known as "post hoc ergo propter hoc." I refer, firstly, to the belief (which I have already mentioned) in the causation of what are called "birth-marks" by "maternal impressions," by which is meant the seeing of unusual and impressive things by the mother when with child; and, secondly, to the belief that a thoroughbred mare can be so affected or infected by the sire (say a zebra) of one foal as to convey to the foal of a later sire (say, a thoroughbred like herself) marks (such as stripes on the legs) which were not present in the second sire, though present in the first sire. This supposed occurrence is called "telegony," and is by some persons supposed to occur in dogs, cattle, and other animals, including man, as well as in the horse.

There is little support in ordinary experience for the belief that birth-marks are caused by maternal impressions, although some of those who are concerned in a professional way with breeding operations cling to it. In very ancient times we find that there was a belief in it, as shown by the story of the patriarch Jacob, who, wishing to obtain the birth of spotted or parti-coloured lambs from a herd of sheep, placed in front of the breeding ewes stakes or rods from which he had removed the bark in rings, so as to make them parti-coloured. He was supposed to have been successful in this way in impressing the visual sense of the maternal ewes with "parti-colouration," and the belief was that they in consequence produced dappled or parti-coloured lambs. The belief, though not general, is widespread among simple folk that such influences can and do act on animals, and it has been, and is by some, similarly held that a human mother may be influenced by surrounding objects, so that if her surroundings are beautiful she will produce a beautiful child. There is absolutely no ground for this belief—based upon experiment. It is merely an unreasoning assumption of "after this, therefore because of this," based upon the incomplete observation of a few accidental cases of vague coincidence and a tenacious clinging to the belief that it is so because it is difficult to prove that it is not so. No trustworthy investigation or experiment on the subject is on record.

But this unwarranted, untested belief, originating among barbarous peoples, has led further, owing to the inveterate love of marvels still common among us, to the notion (surviving to the present day) that the irregular coloured or obscure marks sometimes found on the skin of a child at birth, and vaguely resembling an animal or a fruit, or what not, are due to the mother having recently seen, under some sudden and startling circumstances, the object which the "birth-mark" on the child resembles. Thus we have the following stories related in a recent publication ("Sex Antagonism," by Walter Heape, F.R.S.). The author holds that this strange influence of "maternal impressions" is possible —a matter of comparatively small importance, since the real question is not as to the "possibility" but simply (as in a whole series of beliefs as to more or less improbable occurrences) whether there is or is not sufficient evidence that the connexion and influence believed in actually exists. Mr. Heape relates (without giving any detailed evidence whatever in support of the conclusion which he accepts) the supposed case of a red "mark" like a lizard found on a new-born child's breast being "produced" by the fall of a lizard from the ceiling (the event happened in China) on to its mother's breast shortly before the child's birth. Another case is that of a woman whose husband was brought home from work with his arm lacerated by machinery. Her child was born soon afterwards, and is stated to have had marks on one arm "similar to" those the mother saw on the corresponding arm of her husband. Another story is that of a lady who had a great craving for raspberries before her child was born, and accordingly bore a child with a red raspberry mark on its body!

In no case does Mr. Heape give any picture of the birth-mark and the thing supposed to be represented by it, nor state that he has seen either the mark or a picture of it. In no case is the statement of the mother as to her having been "influenced" as described in the narration, tested or examined in any way.

These and similar stories are related to-day, and such stories have been related from time immemorial. But they are always "hear-say." The witnesses and the facts are never carefully examined, and the degree of closeness of the agreement between the mark and its supposed cause are never really demonstrated. Nor has anyone undertaken a statistical examination with the view of showing that the vague agreement of the mark with the arresting object seen by the mother is anything more than an accidental coincidence, nor (in regard to many such stories) has it been proved that the mother really did see or notice any such terrifying object as she afterwards declares (and possibly thinks) she did. Moreover, no one has carefully and scientifically made crucial experiments with animals, similar to that of the patriarch Jacob. The experiments and their record would not be difficult with animals. Though some farmers may believe that such influences do operate on their breeding dams, there is no known or recognized application of Jacob's method to the production of desired form or colour in domesticated animals. We are not concerned with "possibilities." What is needed is a series of demonstrative experiments, or critical cases. And these are, as yet, not forthcoming.

Telegony is the name given to the hypothesis that the offspring of a known sire sometimes inherit characters from a previous mate of their dam. The name means reproduction (Greek, gonos) influenced by a remote agent (Greek, tele = from afar). There is no question about "possibility" here. Such an "infection" of a dam by a previous mate is not improbable. According to Darwin "farmers in South Brazil are convinced that mares which have once borne mules, when subsequently put to horses, are extremely liable to produce colts striped like a mule." On the other hand, the Baron de Parana states that he has many relatives and friends who have large establishments for the rearing of mules where they obtain from 400 to 1000 mules in a year. In all these establishments, after two or three crossings of the mare and ass, the breeders cause the mare to be put to a horse; yet the pure-bred foals so produced have never in a single case resembled either an ass or a mule.

A celebrated case to which Darwin attached importance was that of Lord Morton's mare, reported to the Royal Society in 1820. This mare, after bearing a hybrid by a quagga (a striped equine related to the zebra) produced, to a black Arabian horse, three foals showing a number of stripes, and in one of them more stripes were present than in the quagga hybrid. This seems at first sight strong evidence in favour of "infection" of the mare by the early quagga mate. But it appears that stripes are frequently seen in high-caste Arab horses, and colts cross-bred from such and other breeds of horse sometimes present far more distinct bars across the legs and other zebra-like markings than were seen in the late offspring of Lord Morton's Arabian mare. The fact appears to be that all the living species of the horse family (horses, asses, quaggas, and zebras) are descended from an ancestry of "striped" equines, and are liable occasionally to "throw back" to their striped ancestry, more or less.

Professor Cossar Ewart determined some years ago to submit the matter to direct experiment, and has related his results in a book ("The Penicuik Experiments," 1899). The South African equine called the quagga, which was that used by Lord Morton, having become extinct, Professor Ewart made use of a richly striped Burchell's zebra. Thirty mares put to this animal produced seventeen hybrids, and subsequently these mares, put to horse-stallions, produced twenty pure-bred foals. All the zebra hybrids were richly and very distinctly striped. Of the twenty later pure-bred horse-foals from the same mares three only presented stripe-like markings at birth, and these were few and indistinct. They disappeared when the foal's coat was shed. Their mothers were Highland mares. But the value of the faint striping in these three instances as evidence in support of telegony is at once destroyed by the fact that Professor Ewart obtained at the same time pure-bred foals from similar Highland mares which had never seen a zebra. Two of these pure-bred Highland foals showed stripes at birth, and one acquired stripes later; and further, whilst the stripes on the foals born after hybrids had been produced by their mothers disappeared with the foal's coat, the stripes on the three pure-bred colts whose mothers had never been near a zebra persisted for a longer period. Similar experiments confirmed these results, showing that traces of striping are no more likely to occur on the offspring of a mare which has previously produced a mule with a zebra or an ass, than on one whose dam has neither seen nor been near to a zebra or an ass. Lord Morton's case thus falls to the ground.

Breeders of dogs are (or were) even more thoroughly convinced of the fact of telegony than breeders of horses. But Sir Everett Millais, who devoted thirty years to the breeding of dogs and experiments on this question, states that he has never seen a case of telegony. And recent experiments of the most definite kind support his conclusion. Dalmatians, deerhounds, and retrievers have been used in these experiments. Many such experiments in telegony are accidentally or unwittingly made every year with dogs. An undesired crossing of two breeds takes place, but when subsequent pure breeding takes place no "telegonic" infection of the mother is observed. Cases believed to be due to telegony have on examination proved to be due to the carelessness of stablemen, who have allowed a dog to escape temporarily from the kennels or to enter them uninvited. The men have attributed the mongrels so begotten to telegony in order to conceal their negligence.

Another curious case was that of a rickety spaniel puppy, which was exhibited a few years ago at the Zoological Society and believed by the exhibitor to owe its bandy legs to "telegonic" infection of the mother by a dachshund, with which she was supposed to have mated a year or more before being put to the father of the spaniel. Its true nature was at once recognized by the experts present, the bandy legs being those caused by "rickets," and not like those of the well-known dachshund breed.

It appears that the explanations widely prevalent of many apparently strange things discussed in the preceding chapters, such as live toads buried in rocks, the water-finder's mystic rod, the coincidence of birth-marks and maternal impressions, and the inheritance of offspring from a previous mate of their dam, are hasty and unverified suppositions, which have never been properly tested, and that when the wonder-provoking statements made and the actual facts in question are properly and sufficiently examined, according to the rules of evidence and common sense, it is discovered that the assumption of occult or exceptional causes in explanation of such strange things are not justified, but that these strange things owe their strangeness in large part to the incorrect and incomplete observation of those who report them, and to that love of marvel and mystery which, like hope, springs eternal in the human breast.

It is a remarkable proof of the reality of the belief

in telegony—though not a proof of the reality of telegony—that amongst breeders of horses and dogs the selling value of a dam which has borne young to an inferior sire or to one of a distinct species, is largely diminished as compared with that of a dam which has been mated with a first-rate sire of her own breed. Darwin himself was led, by his inquiries into a similar occurrence in plants, to favour the notion that a sire could so "infect" a mare that her offspring by a later sire would in some instances show traces of the characters of the earlier sire. The parts of a plant which form the coverings of the fertilized ovule, the "coats" of the seed and the seed-case and fruit, are, of course, parts of the maternal plant. In each of the ovules which grow in the central part of a flower (the so-called "pistil") is an egg cell like that of an animal. This is "fertilized" by the pollen-grains which are brought by wind or by insects from the "stamens" of another flower. Each pollen-grain thus brought to the surface of the pistil elongates into a delicate filament, and penetrates into it, and so reaches an egg cell, with which it fuses. Then the surrounding tissues grow and swell up, forming the seed coats and the fruit. They are parts of the egg-cell-producing or "mother" flower. Thus the pulp and "rind" or skin of an orange is part of the mother plant, not of the germs or young embedded in the "pips." It is found that if an orange-flower is deliberately fertilized by placing on its pistil the pollen-grains of a lemon-flower, not only are the ovules of the orange fertilized, but the surrounding structures, which enlarge to form the fruit and are parts of the orange plant quite distinct from the ovules, also become affected by the pollen. In one well-observed case when an orange-flower was fertilized by a gardener with the pollen of a lemon-flower, the skin or rind of the resulting fruit was found to exhibit stripes of perfectly characterized lemon peel (having the colour and flavour of lemon peel), alternating with stripes of the proper orange peel.

The same thing has been observed in apples, melons, orchids, rhododendrons, grapes, maize, and peas, when one variety has been fertilized by the pollen of another, or when one species has been fertilized by the pollen of an allied but distinct species. The fruit in these cases (not simply the germ or young plant within it) has been found in some instances to have some of the colour, flavour, or shape and marking of the fertilizing variety or species blended or else mixed like a patchwork with that characteristic of the fertilized variety or species. The egg-producing or mother plant not merely has its ovules fertilized, but its tissues for some distance around are infected and made to take on—in parts of their living, growing substance—some of the quality of the fertilizing species. A similar thing occurs, though rarely, when cuttings of one plant are grafted on to another. The living tissue either of graft or of stock, and sometimes of both, is affected by the fusion with it of the tissue of the second plant united with it. And this appears to be a kind of "infection"—living particles passing from one to the other, and producing a mosaic or patchwork of the two kinds of living substance characteristic of each of the united plants.

If an individual flower were to produce in a second year after its first fertilization and seed production a second set of ovules which could be fertilized by a kind of pollen differing from the first, it would not be surprising did that second set of ovules sometimes show characteristics due to the infection of the maternal tissues by the pollen used in the first year. But flowers do not survive and produce ovules in a second year. They are completely used up each year, and drop off as "fruits" from the plant which bears them. With many animals, however, the facts are otherwise. The same mother produces from the ovary year after year successive ovules, and it would thus be quite intelligible that the fertilizing sperm of one year should frequently have so affected or infected the egg-producing organ or ovary as to result in the conveyance to the later crop of egg cells separated from the ovary, some of the qualities of the earlier male parent. These considerations warrant the guess or "hypothesis" of telegony in animals. But all such guesses must be put to the proof, and not accepted simply because there is no reason to conclude that they are impossible. As things at present stand, there is no evidence, resulting either from deliberate experiment or from exact observation and record of the natural breeding of animals, to justify us in holding, as an established fact, that the offspring of a given sire and dam is, even in rare cases, affected by the previous mating of the dam with another sire. Naturalists would be deeply interested in the production of even one indisputable instance of this occurrence.

In connexion with this matter it is to be noted that the sperm of one drone (her only mate) is retained in an internal sac or pouch, alive and active, in the queen bee, for some four or five years, and is used by her in successive seasons for fertilizing her eggs. Similarly it is recorded by the late Lord Avery that a queen ant kept by him for fourteen years, without access to a male ant, retained to the end of that period the power of producing eggs which developed into worker ants. He concluded that the sperm received fourteen years before by this queen from a male ant remained all this time alive and ready for use in her sperm-receptacle or sac, since it has been shown that unfertilized eggs in these and allied insects produce only drones (males).

Many strange and unwarranted beliefs persist because mankind prefers to accept an astonishing assertion as true rather than take the trouble to see whether it is so or not. Thus all antiquity and the later learned world wrangled about the very existence of Homer's city of Troy, until Schliemann said, "Don't talk! Dig!" and with childlike simplicity and directness uncovered ancient Troy. Thus the belief as to St. Swithin and his forty days of rain has been shown by the simple examination of the actual records of rainfall to be very far from the truth, since, though we often have a wet period in July and August, St. Swithin's Day is nearly as often free from rain in a wet season as the reverse. Forty days of rain very rarely indeed, in the South of England, have followed a wet St. Swithin's Day. The most amusing instance of the pricking of one of these bubbles of belief arose from the inquiry by some of the sham philosophers at the Court of King Charles II as to how it comes about that if a jar holding water be weighed, and then a live fish be placed therein without spilling any of the water, and the jar, with the fish and the water in it, be again weighed, there is found to be no increase in the observed weight. King Charles, it is said, made a bet that this was not so, and that there was nothing to explain. He referred the matter for decision to the newly founded "Royal Society for the Promotion of Natural Knowledge," which at other times he had asked to give him information as to the magic properties of the unicorn's horn and the cause of the movements of the recently imported "sensitive or humble plant." The believers in the marvellous disappearance of the weight of a fish placed in a bowl of water held forth at great length and gave ingenious reasons as to why this is so. But the King said, "Don't chatter; make trial!" And the weighing was done, in the King's presence, by some of the Fellows of the Royal Society. It was found that the weight of the jar with its contained water was increased when the fish was placed therein by exactly the number of ounces which the fish weighed when placed separately in the balance. So the King won his bet, and the sham philosophers were silenced. The whole spirit of science, as contrasted with that of superstition and ignorance, is summed up by the Royal Society's motto, "Nullius in verba" (on no man's assertion!), and the King's command, "Don't chatter; make trial!"

CHAPTER XLI

HOW TO PROMOTE SCIENTIFIC DISCOVERY BY MONEY

THE fact that five years ago Mr. Otto Beit, the brother of the late Mr. Alfred Beit, not only carried out the latter's intention of giving £50,000 to the promotion of research in connexion with the study of disease and the mastery of its causes, but added £150,000 on his own account to the amount originally proposed, produced great satisfaction among scientific men, and also in that large body of the public which, at the present day, understands something of the importance to the community of the minute and thorough study of disease, of its mode of access to man, and of the possibilities, which every day become brighter and clearer, of getting rid of it altogether. All honour and gratitude are due to Mr. Beit for his generous gift and for his wise appreciation of the good which can be done by proper application of such a fund. I have reason to know and to value the large-minded interest in science which was shown by the late Mr. Alfred Beit, since he gave me £1000, some twelve years ago, towards the expenses of expeditions which I was organizing for the investigation of the natural history of Lake Tanganyika,—expeditions which have yielded important scientific results, and have but recently exhausted the fund then collected.

It has often occurred to me that wealthy men who wish to devote large sums of money to the promotion of scientific research find difficulty in carrying out their intentions, owing to the fact that they do not know enough about the methods and conditions of scientific discovery to enable them to form a definite independent judgment as to how to assign their money, so as to make sure that it shall really be employed in the most effective way towards the end they have in view—namely, the increase of scientific discovery. They naturally have some doubts as to whether the old (or even the new) Universities can help them as trustees of the money when they see the importance attached by the former to antiquated methods of teaching and examination and observe their traditional cultivation of certain favoured studies, with a minimum of activity in research and discovery. They mistrust special societies or individuals as advisers in the matter, and sometimes finally spend the money which they had destined to be the means of furthering scientific discovery upon a costly and ill-considered architectural monstrosity dedicated to science, but of little help to its progress.

In past times various schemes have been adopted by benevolent men for bequeathing or giving their money so as to promote scientific discovery. Very generally there has been a certain amount of confusion between two distinct purposes—namely, that of creating new knowledge (the discovery of previously unknown things and new processes), and that of spreading existing knowledge amongst an increased proportion of the community. An admirable endowment for the latter purpose is that of Mr. Smithson, a member of the family of the present Duke of Northumberland, which was refused by the British Government for peculiar reasons, and conveyed by that gentleman to trustees in the United States of America about a hundred years ago, where the Smithsonian Institution has vastly aided the spread of science. Another valuable endowment which has been administered by special trustees for a still longer period is that of the celebrated physician Radcliffe, to whom we owe the scientific and medical library, an astronomical observatory, and travelling fellowships in the University of Oxford. The greatest sum dedicated to scientific research in England of late years is the noble gift of a quarter of a million sterling made by Lord Iveagh to the Lister Institute of Preventive Medicine. There have been not a few generous donors of smaller sums for like purposes.

An inquiry was set on foot a few years ago in America in order to obtain the opinions of those who had experience of scientific research and the institutions intended to promote it in different countries, as to the best methods to adopt in order to effect such promotion. I do not know whether any report was published, but I remember that I was consulted on the subject by the late Professor Simon Newcomb, a foreign member of the Royal Society and one of the most distinguished scientific discoverers in the United States. I am quite sure that no general agreement or conclusion on the subject has been arrived at. So far as I can see, whenever any high-minded philanthropist desires to devote in this country a large sum of money to the promotion of scientific discovery, he is liable to come under the influence of highly respectable and eminent persons who, although they have no acquaintance with the nature of scientific discovery and the way in which it actually takes place, do not hesitate to fix up a scheme based on some antiquated and mistaken model, which is accepted with simple faith by the benevolent donor.

Scientific research is a delicate plant, and the secret of the way in which it may be nurtured has not been revealed to dignitaries and officials. It is interesting to note some of the methods which have been tried with the object of nurturing scientific discovery. In every case the donor has chosen or created an electing body or trustees of which I will say more below. He has directed this body to expend his gift with a view to the promotion of scientific discovery in one of the following ways: (1) in awarding prizes for discoveries made; (2) in terminable stipends to junior and senior workers selected by the trustees and called scholars or fellows, the stipends being given on condition of their holders devoting themselves for a few years to the attempt to make discoveries; (3) in permanent salaries to tried men, who are thus paid as professors or directors of laboratories and museums; (4) in providing specially designed buildings and apparatus for research, but no salaries for the workers; (5) in providing, on whatever scale the fund given permits, groups consisting of a professor or director, two or more assistants, attendants, building, apparatus, and the annual income necessary for materials of investigation and maintenance of the establishment. As to the trustees, or boards of electors, chosen by the donor, they are often some established scientific society or some university, or the board may be specially appointed by him. The last is the best sort of body, if properly constituted, but not unfrequently the perplexed promoter of scientific discovery finds himself assenting to the constitution of what is called "a representative body"—say, a bishop, a town councillor, a Secretary of State, a judge, and a university professor, with other members to be nominated by himself or his heirs. Such a board fails from a want of knowledge.

The methods of applying the income provided by the donor are not always such as to produce any marked result in the direction desired by him. It is generally agreed among scientific workers and experts that the giving of prizes or rewards for scientific discovery does not tend to increase the output of discoveries, however carefully and justly awarded. Though such an award as the £8000 or £10,000 of the Nobel prizes is a very agreeable compliment to the man so honoured, and often richly deserved, no one would urge a would-be promoter of scientific discovery to devote his gift to the foundation of prizes. And so, too, with regard to scholarships or fellowships, it is very generally and rightly held that they do little or nothing in promoting scientific discovery when they are small in value and are only to be held for two or three years. When a young man has taken his university degree in science or medicine a scholarship or fellowship of £250 a year for three years offers no inducement to him, if he is an able man, to abandon his regular professional career. If he accepts it, he will have had no time to go far on the path of discovery before it comes to an end, and he will find at the end of his three years that he has lost that amount of time so far as his profession is concerned, and that there is no life post or career open to him in the line in which he has spent three years—namely, that of a scientific investigator. As a rule, able men will not be drawn off in this way from their professions, but inferior men may be.

The man, on the other hand, who is specially gifted with the power of scientific discovery will not be affected by such temporary fellowships. He will enter on the career of discovery with or without such inducements. What such a man (and he is the only sort of man who matters) really requires, and should find open to him, is an assured career. This must take the form in the first place of a smaller post as assistant to a great discoverer, tenable for twenty years if need be, and subsequently a life post, with laboratory and assistants, when he has proved his possession of the discoverer's quality. Hence it is that what the benevolent millionaire who wishes to promote scientific discovery should do is to provide life posts, "professorships" or "directorships," for the really great discoverers, who exist often in cramped conditions. They should be of the value of £1500 to £3000 a year—not too large a stipend in view of the incomes earned by successful professional men and assigned by Government to judges, bishops, colonial governors, senior civil servants, and politicians—with two or three assistantships of £150 to £500 a year attached, to be filled up by nominations made by the professor himself as vacancies occur. A sum of £7500 a year, that which Mr. Otto Beit has so generously given, would pay for one professor, with three assistants, attendants, and interest on building and maintenance fund. Of course, if such a sum were offered to an existing institution where buildings and other conveniences are already provided, two research professors and their assistants could be paid for where one only would be possible if building and service had to be provided. There are buildings and laboratories in London and elsewhere provided by beneficent founders without stipends for directors and assistants, and there are already a good many young graduates drawing terminable inadequate stipends in succession to one another from great foundations. The difficulty is to bring about the combination of adequate funds for the chief and for the graduated minor posts, and for a well-equipped laboratory. When that is done, as it sometimes, though rarely, is, the only further difficulty is how to choose a real man, an inspired, inspiring discoverer. There is only one way.

Real discoverers are extremely rare—great ones are recognized about once in fifty years in any one large branch of science. There may be others wandering about—undiscovered discoverers. The only people who can discover them are men like themselves. Hence, in German universities and all wisely managed institutions for the promotion of scientific discovery, they give the power of choosing new discoverers to those discoverers already belonging to the university or institution, and they take care that all the electors are vitally interested for the honour, credit, and pecuniary success of their university. These conditions can be arranged and brought into healthy action by care and understanding. But the whole fabric may go to pieces, and jobbery and jealousy prevail (as has sometimes happened in England) if care is not taken to identify the personal interests of the electors (brother professors) with the honest exercise of their capacity to choose a real discoverer to fill a vacancy when it occurs, or if an ignorant council of "superior persons" is allowed to interfere.

To find these great discoverers is, indeed, no light task. They have to be looked for by the State, firstly, in the primary schools; the net has to be drawn and the minor fishes allowed to escape, whilst the strong and promising are sent on to high schools. Then again, after further sifting, some are passed on to the special college, then a selection to the university, and at last one or two a year may be chosen as assistants to an established and inspiring discoverer. Seven, ten, or fifteen years later one out of all his fellows and predecessors is recognized as the incomparable teacher and discoverer—the inspirer of others, the one great man of half a century. He must be chosen by his colleagues, his fellow-workers, not by political wire-pullers nor by any variety of social "Bumble." He is given laboratories and assistants, and men come to consult him, to sit under him, work for him, from all parts of the world. Louis Pasteur was such a man. Huxley pointed out by what a vast public expenditure Pasteur was gradually sifted out from his fellows, and made professor in the Normal School of Paris. Of course, a good many inferior people got a share of the training provided, and did some unimportant things; but if we put them aside it is perfectly true (as a calculation of the expenses of the whole network of State-supported schools and colleges and bursaries through which he passed will show) that the capture or discovery of Pasteur cost the French nation about £25,000,000. He was worth it, not only to France, but to every other nationality—and more, too, more than can be measured by gold. His name, honoured throughout the world on account of the splendid discoveries associated with it, gave self-respect, courage, and healthy pride to France at a time when she had cruelly suffered. Ten years ago the most popular newspaper in France took a "plebiscite" to determine who, in the general estimation of the French people, was the greatest Frenchman of the nineteenth century—the century which included the first Napoleon, Victor Hugo, Gambetta. The vote was given by some millions, and resulted in a majority for Louis Pasteur. Would Englishmen have shown such discernment? Such a man is absolutely necessary as the head of any great institute which exists for the purpose of scientific discovery. Such men, smaller it may be, but of the same inspiring quality, are the only men fit to be university professors. It is because there are still such men at the Institut Pasteur that it remains a great seat of discovery. It is because they have not such men, and that there is no intelligent attempt to get them, that many wealthy institutions in our own country fail to produce scientific fruit.

INDEX